The Climate of
EDMONTON
by Rod Olson

**Climatological Studies
Number 37
The Climate of Canadian Cities
Number 2**

A Publication of the Canadian Climate Program

Catalogue No. En57-7/37E   Canada: $6.95
ISBN 0-660-11936-6   Other Countries: $8.35

Price subject to change without notice

THE CLIMATE OF EDMONTON

SUMMARY

Edmonton has a northern mid-latitude continental climate. As such, the summers are warm, the winters are cold, and most of the precipitation falls during the summer months. Variability is the dominant characteristic of Edmonton's weather and climate, and large departures from the long-range average can be expected from day to day, month to month, and year to year. The general climate is described in terms of two main seasons, summer and winter, and two transition periods, spring and fall.

Applied climate is discussed under the categories of comfort, winter heating, summer cooling, recreation-tourism and gardening.

SOMMAIRE

Le climat continental d'Edmonton est celui des latitudes septentrionales moyennes, avec des étés chauds où se produisent la plupart des précipitations, et des hivers froids. La caractéristique dominante du temps et du climat d'Edmonton est la variabilité, et l'on peut s'attendre à des écarts importants par rapport aux moyennes à long terme, d'un jour à l'autre, d'un mois à l'autre ou d'une année à l'autre. Sur un plan général, nous distinguons deux saisons principales, l'été et l'hiver, et deux périodes de transition, le printemps et l'automme.

Nous étudions les effets du climat sous l'aspect du bien-être, du chauffage en hiver, de la climatisation en été, des loisirs et du tourisme ainsi que du jardinage.

TABLE OF CONTENTS

Page

viii

LIST OF FIGURES

Page

ACKNOWLEDGEMENTS

Mr. A.S. Mann, Atmospheric Environment Service (AES), Edmonton, inspired this work. He initiated the project and provided valuable suggestions and guidance through its entirety.

Mr. R.B. Thomson, AES, Edmonton, was the sounding board for many of the ideas. He reviewed the manuscripts and his comments were always so timely.

Members of Scientific Services Division, AES, Edmonton, provided assistance whenever requested — Messrs. P.J. Kociuba, W. Prusak, H. Turchanski, E. Hawthorne, J.P. Ross, M.D. Drews and M. Cote.

Editorial and publishing services were supervised by Mr. R.B. Crowe at the Canadian Climate Centre, Downsview, Ontario.

INTRODUCTION

The City of Edmonton, situated along the North Saskatchewan River in central Alberta, is home to over a half million people. It is located within the transition zone between the prairie grasslands and the northern forests. It usually experiences the warm summers of the prairies and the cold winters of the boreal forest. Variability is a dominant characteristic of the climate within this region.

In Edmonton, as with most centres, the climate and weather are frequently the topics of discussion and most people have their own perceptions about them. For example, on 21 April 1883 the editor of the <u>Edmonton Bulletin</u> wrote:

"In the north the constitution of men...is built up and strengthened by nature in order to prepare for the cold winters...while in the south where the ordinary weather of winter is not so cold, no such strengthening process takes place ... The people of the North then are more healthy, more happy and more progressive...than those of the South."

More than a hundred years later a similar sentiment endures. While reflecting upon the passing of the seasons the editor of the <u>Edmonton Journal</u> (2 August 1983) mused that the character of Edmontonians

"...is drawn from the land.... [They] appreciate the rhythm of nature here, which becomes the rhythm of the people, too".

Climate and weather are often misinterpreted as being synonymous. Weather is the actual day-to-day condition of the atmosphere. Climate is the long-term manifestation of weather and it brings together the history of day-to-day weather in all its forms - temperature, wind, sunshine, cloudiness and precipitation.

Historic climate data for Fort Edmonton, Northwest Territories go back to 29 December 1879 when G.S. Wood, a telegrapher, installed the first meteorological station in the City. Daily observations commenced on 11 July 1880 (Table 1) with minor interruptions when Mr. Wood was repairing the telegraph line. In 1882 the site was moved to 115th Street and Jasper Avenue, and

TABLE 1

**Length of temperature (T) and precipitation (P) records
for stations in the Edmonton area**

	T	P		T	P
1. Bremner		1962-1972	11. Edmonton Namao A	1955-	1955-
2. Clover Bar		1906-1907	(continued)		
3. Edmonton	1880-1882	1880-1882	12. Edmonton Pleasantview		1977-1979
4. Edmonton	1882-1912	1882-1912	13. Edmonton University	1919-1919	1919-1919
5. Edmonton	1912-1943	1912-1943	14. Edmonton Woodbend	1973-	1973-
6. Edmonton Calder	1975-1977	1975-1977	15. Ellerslie	1964-	1964-
7. Edmonton Coronation	1978-1979	1978-1979	16. Fort Edmonton	1976-	1976-
8. Edmonton Int'l A	1960-	1960-	17. Fort Saskatchewan	1958-	1958-
9. Edmonton Lynnwood		1971-1976	18. Jasper Place School	1962-1966	1962-1966
10. Edmonton Municipal A	1937-	1937-	19. North Cooking Lake	1918-1930	1918-1930
11. Edmonton Namao A	1944-1945	1944-1945	20. Strathcona	1925-1927	1925-1927
(continue next column)			21. Winterburn Lakeshore A	1964-1964	1964-1964

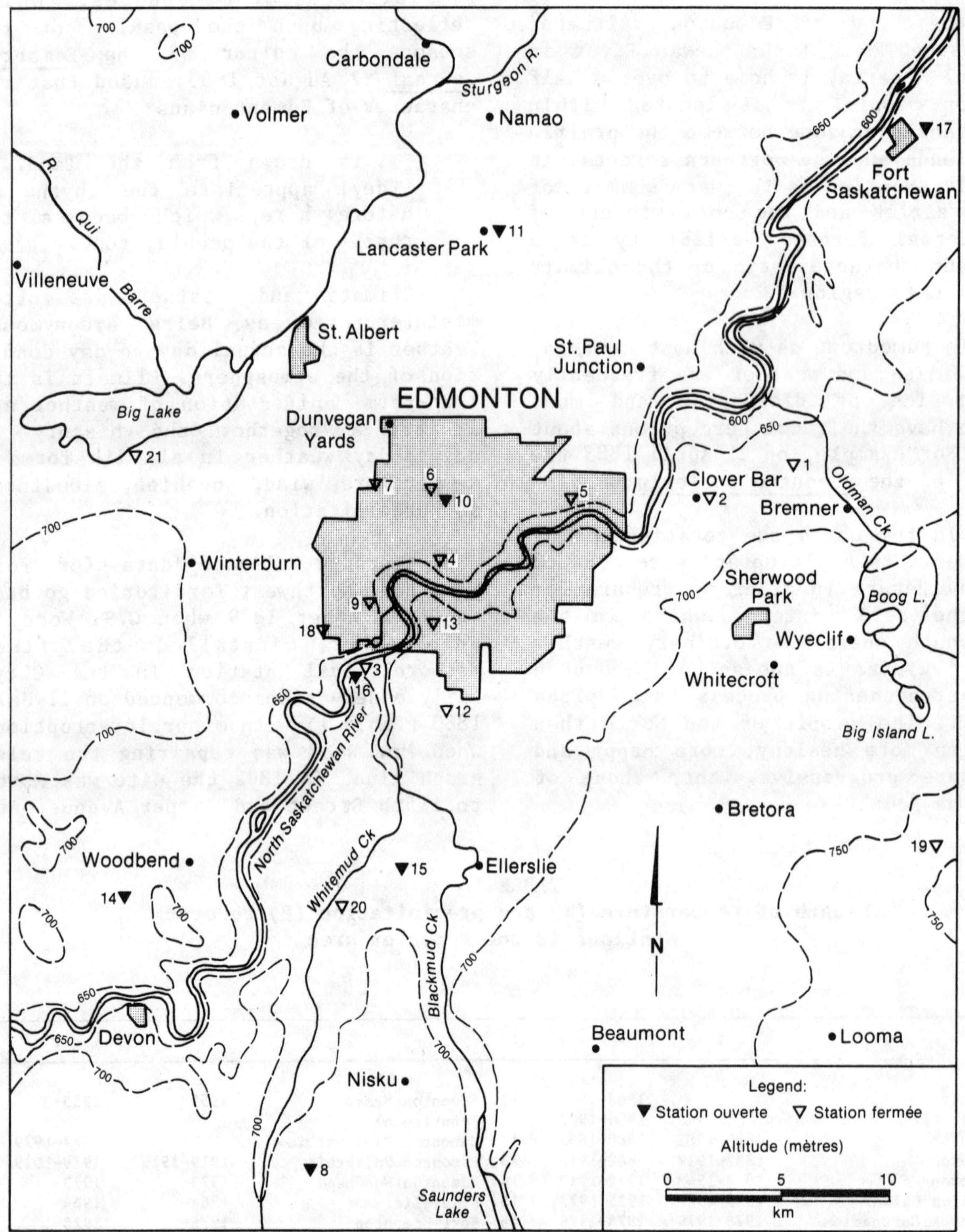

Figure 1. Topographical and physical features and location of climatological stations in the Edmonton area

the responsibility for the meteorological station was delegated to the Hudson Bay Company. The site was moved in 1912, this time to 63rd Street in the Highlands District. Observations were continued there until the station was closed in 1942. On 1 September 1937 a meteorological office operated by the Federal Government opened at Edmonton Municipal Airport and it became the official reporting station for the City.

In more recent times three other major weather observing stations have been established in the Edmonton area. The station at Canadian Forces Base Namao was opened in 1955, the Edmonton International Airport in 1960, and in 1964 a climate station was started at the University of Alberta Farm near Ellerslie (Figure 1). The data from these sites help to define the variability of climate between the City and its surrounding area.

CLIMATE CONTROLS

There are controlling factors that dictate the characteristics of each particular climate. These controls include latitude, topography, elevation, distance from a large body of water, and the circulation of the atmosphere.

Latitude determines both the duration of daylight and the angle at which the sun's energy will be intercepted at the earth's surface. The summer solstice, which is in the third week of June, is the longest day of the year in the northern hemisphere. Edmonton has 17 hours of sunlight on this day and the sun, at 60° above the horizon, is at its highest point in the noonday sky. In winter the sun is low in the sky even at noon. On the winter solstice, which is in the third week of December, it reaches a maximum of only 13° above the southern horizon. The sun

does not provide much heat during this period due to its low angle and shortness of day length.

Edmonton's climate is described as continental and as such, is subject to large contrasts in temperature between day and night or winter and summer. This is because the City is a long way from the ameliorating influence of the Pacific Ocean. In addition, the Western Cordillera acts as a barrier to the flow of air, making the "effective" distance even greater.

Edmonton is in the zone of the westerlies, a large-scale atmospheric circulation system that flows in a general west-to-east direction. These winds entrain and transport vast bodies of air that have developed the characteristic properties of the surface that they had previously overlain - the cool, moist Pacific Ocean; the cold, dry Arctic; or the warm, dry American southwest. The weather in Edmonton on any particular day is largely determined by where the overlying air mass originated. When air masses from contrasting source regions meet they do not mix readily and a front is formed. A front passing over Edmonton will usually produce a change in temperature, humidity, cloudiness, wind speed, and wind direction.

There are many aspects of climate that differ between Edmonton and its neighboring countryside. The metropolitan area exerts an ever-increasing influence on its climate as it expands. The climate elements that are affected include temperature, humidity, visibility, solar radiation, wind and precipitation. Generally, the urban area is slightly milder and less humid and windy than the surrounding countryside. These effects are great enough to produce noticeable differences in the impact of the climate on gardening and transportation between the two areas.

THE SEASONS

A season is a distinct portion of the year with uniform climate characteristics. The City of Edmonton has two periods in which climate conditions are relatively uniform - winter and summer. Spring and fall in the seasonal context are not distinct in central Alberta since their weather on any given day is often similar to either summer or winter. Rarely is there an orderly transition from summer to winter or winter to summer.

The winter-summer duality of Edmonton's seasonal regime can be demonstrated using various climate parameters. Temperature (Figure 2) has two distinct periods, one warm, the other cold, that have relatively uniform conditions. They are separated by short transitions in which temperatures are rapidly increasing or decreasing.

The distribution of total monthly precipitation in Figure 2 illustrates a relatively wet period during the warmer months and a dry period during the cooler months. Associated with the transition seasons, spring and fall, are the year's largest variabilities in precipitation. Often the monthly totals will be very high or very low compared to the long-term normals. Consequently, in any particular year, these months

Fort Edmonton in 1886, looking west from the south side of the North Saskatchewan River. The present day Legislative Building occupies the site of the former Fort (Provincial Archives of Alberta).

can often be grouped with either the wet or dry periods of the year. The occurrence of snow can also be used to divide the year. The warmer five months rarely have measurable snowfalls. The cooler five receive nearly all their precipitation as snow. The remaining two months, April and October, average half their totals as snow.

Winter in Edmonton is characterized by a persistent snow cover. Outdoor activities that require a blanket of snow or a sheet of ice can be conducted throughout this period. The beginning of winter is defined as the average date when the mean daily maximum temperature dips below freezing, which is normally November 16. From this date on, snow that falls will usually stay for the season. Winter continues until the mean daily maximum temperature rises to the thawing point, usually March 17. Snow remains on the ground for a period after this date, but winter activities cease due to patchy snow cover and melting ice.

Summer in Edmonton is characterized by warm weather. Outdoor events are not limited by temperature, but occasionally rain or wet ground can restrict these activities. The season is considered here to begin when the mean

The central business district of Edmonton, looking west (1983). MacDonald Hotel, which was the landmark of the City's skyline until the early 1960's, is almost lost against the backdrop of today's highrise buildings (R. Olson).

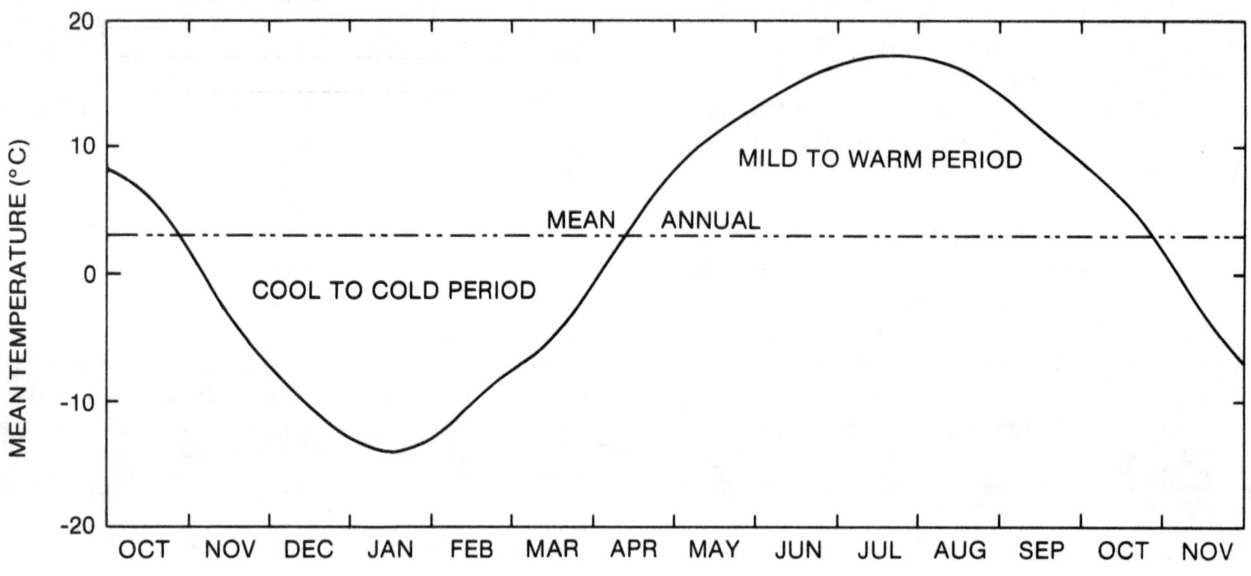

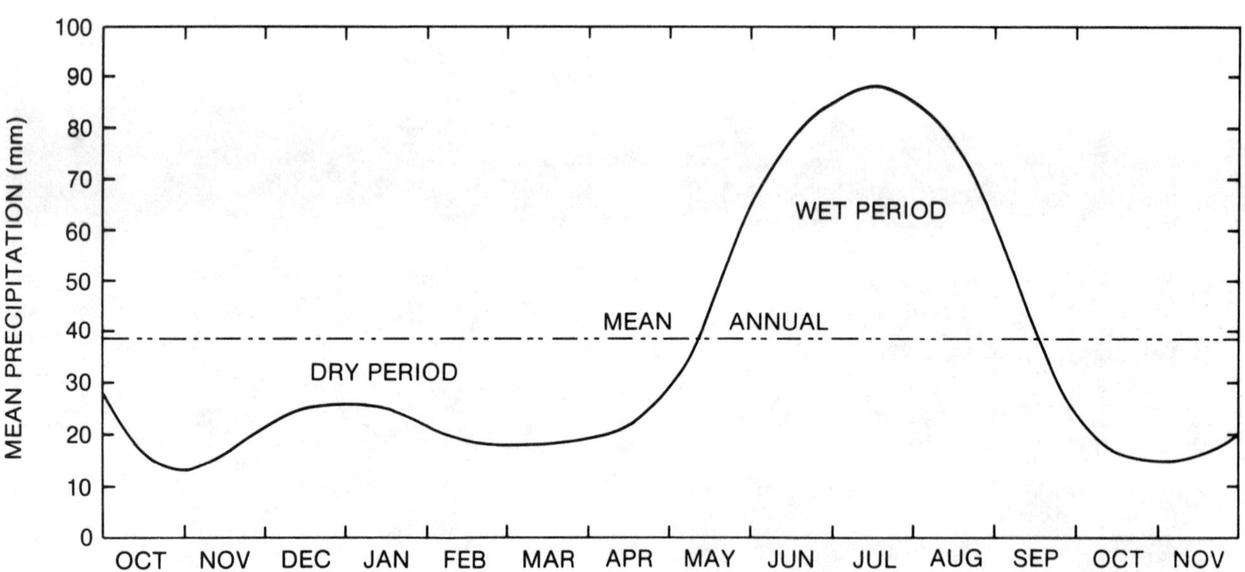

Figure 2. Temperature and precipitation regimes

daily maximum temperature reaches the 18°C threshold, May 19, and it ends the day before the temperature drops below this value, September 6. By coincidence, the beginning and ending of summer correspond to Canada's Victoria Day and Labor Day holiday weekends.

Spring and fall are defined as the periods when mean daily maximum temperatures are at or above 0°C but less than 18°C. Any given day during these transitional seasons may have weather that is characteristic of summer or winter. The closer the date is to the beginning or ending of summer or winter, the greater the probability that the transition period will have conditions similar to the corresponding season. Constraints on most outdoor activities can be expected during these intervals, especially during the spring transition. In late March and early April the snow and ice, if there is

any, will not be suitable for winter recreation. Yet, the presence of the melting snow pack inhibits dry-land activities. Golf courses and unimproved hiking or jogging trails along the river will not be satisfactory for use. During the fall transition many of summer's more vigorous activities can be conducted. Dry weather during this period promotes these activities until the arrival of winter's permanent snow cover.

The relationship between calendar time and seasons is depicted in Figure 3. The concept of two dominant seasons separated by transition periods can be visualized in the figure. Winter and summer are both relatively long compared to the two transitions. Winter averages 121 days in length while summer averages 111 days. The spring and fall transitions are brief, at 63 and 70 days, respectively.

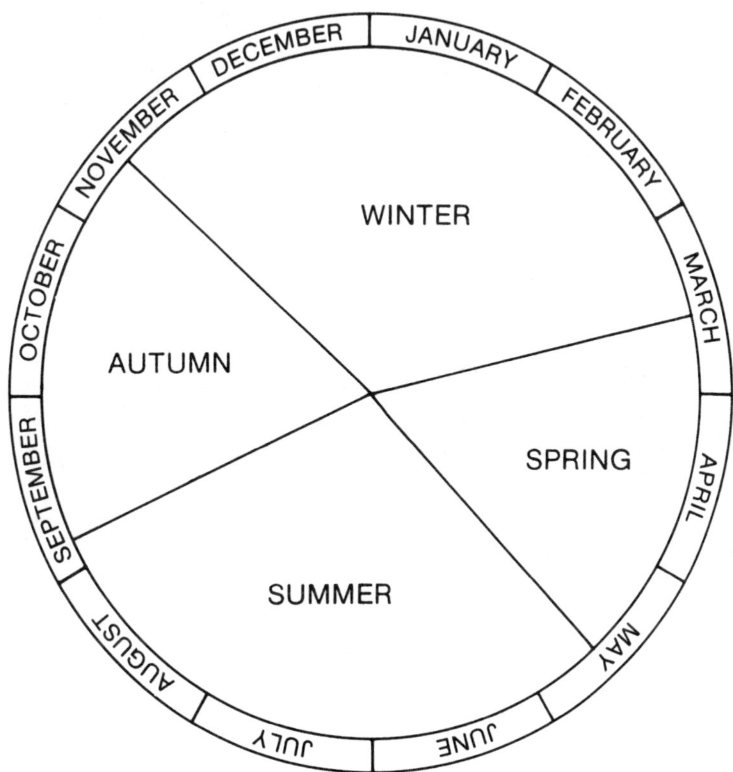

Figure 3. Schematic representation of season lengths in Edmonton

WINTER

The sun's effect as a major climate control is most evident during the winter season. Its noontime position above the horizon rapidly declines during early winter and the daylight period quickly shortens. Consequently, between mid-November and mid-January mean daily temperatures decrease at a rate of about 1°C every four days.

Temperature

The daily temperatures for the winter season are plotted in Figure 4. The mean throughout the period is -10.6°C, but there is more variability in winter temperatures than is the case for any other season. The coldest winter on record occurred in 1886/87 with a mean daily low of -26.2°C. The mildest was in 1888/89 with an average high that was just below freezing. The Edmonton Bulletin reported that fields were being worked on New Year's Eve of that winter. The robins returned the first week of February, the same time that the leaves on the poplars appeared.

Ninety-five per cent of all winters reach -30°C one or more times and two-thirds plunge to -35°C at least once. January is normally the coldest month of the year. On average, half of its nights and a third of those in December and February dip to at least

The Province of Alberta's Legislative Building, Edmonton (1940, A. Blyth Collection, Provincial Archives of Alberta).

-20°C. January usually has one night with a temperature of -35°C, while December or February reach this value in one winter in ten.

A classic chinook, where westerly winds strengthen and temperatures reach the upper teens, rarely occurs in Edmonton. This is due to the distance from Edmonton to the Rocky Mountains. Nevertheless, mild Pacific air almost always reaches the City at least once during each winter month. January averages seven mild days with temperatures above freezing. Every 20 years there will usually be one January that does not have a thaw and one with half of its days above 0°C. There have only been two Februarys that did not reach the melting point, but in February 1977 every day reached or exceeded 0°C. In most years the extreme monthly maximum during the winter will rise to about 6 or 8°C.

Ice

Each winter the North Saskatchewan River at Edmonton freezes over. Ice will first begin to grow along the shore, followed by the formation of ice pans which become more consolidated as November progresses. In an average year the surface ice ceases movement on 16 November, the first day of a typical winter. The earliest that shore ice

Winter in Edmonton along the banks of the North Saskatchewan River. The Hotel MacDonald dominates the skyline of the central business district (1942, A. Blyth, Provincial Archives of Alberta).

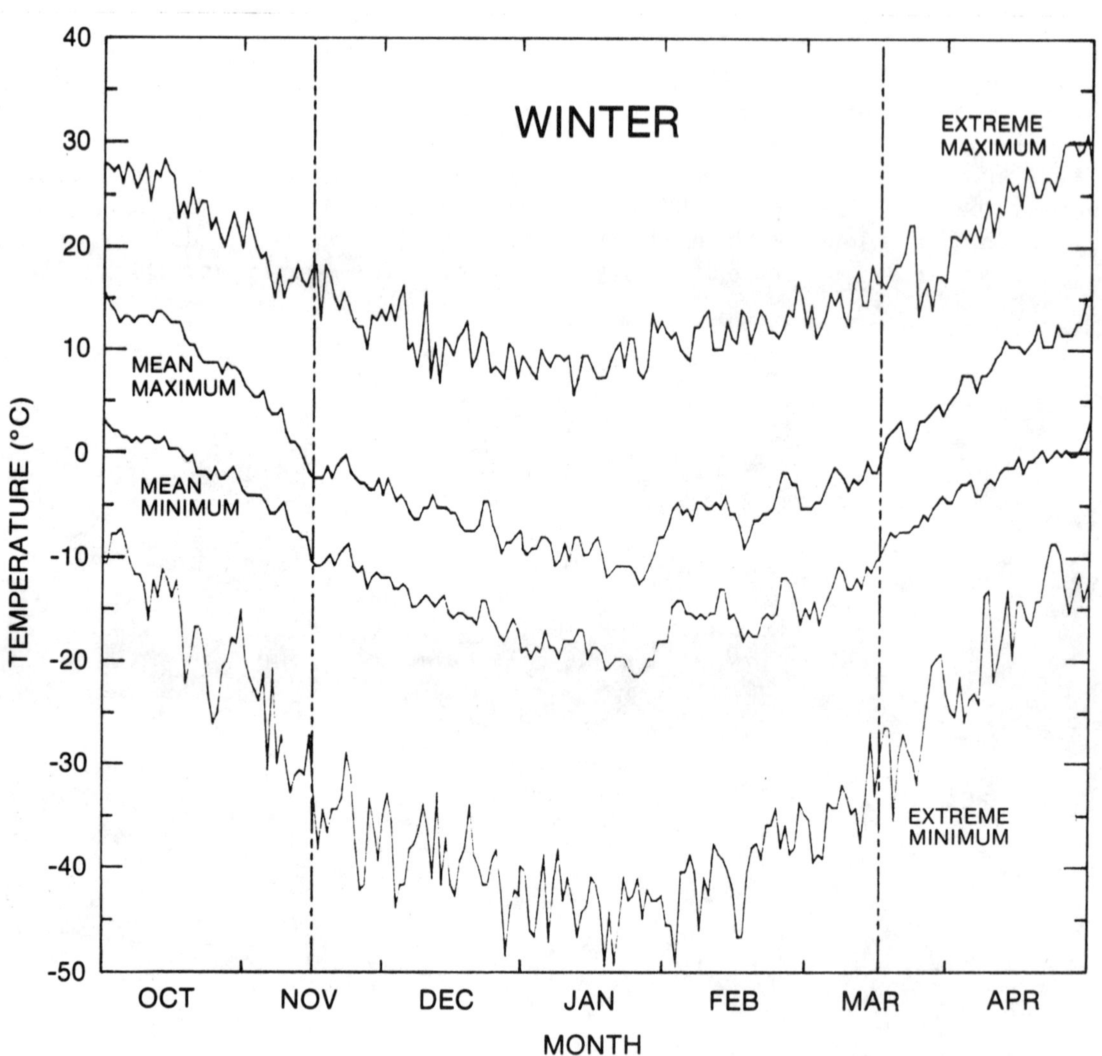

Figure 4. Mean daily and extreme temperatures during winter

began to form on the river was 11 October 1881, the latest was 24 November 1962. The earliest complete freeze-over was 23 October 1881 while the latest was 15 December 1962.

Precipitation

Winter is the longest of the seasons in Edmonton, yet it only accounts for 18% of the year's total precipitation. An average winter will receive 93 cm of snow. This amount has a water equivalent of 86.2 mm, which is less than the mean precipitation during July alone. The snowiest December-February was 1903/04 when 150 cm fell. Including the snow that fell before and after the three-month defined winter, there was almost 200 cm recorded. Two winters later, between October 1906 and May 1907, a total of 239 cm fell. The winter of 1888/89 had the least snow. There was only 15.6 cm throughout the season and it melted soon after falling.

An average winter in Edmonton will have approximately one day in three with measurable precipiation. The winter of 1888/89 had three, while the winters of 1964/65 and 1973/74 had 62 each. Each winter will usually have one storm that deposits up to 9 cm, while a snowfall of 10 cm or more occurs about once every five years. The greatest single snowstorm during winter proper was in December 1979 and it deposited 25.8 cm.

In general, January will have more snow on the ground than any other month. By the end of this month there will usually be a blanket of 23 cm. December averages 15 cm at month-end, but about once every 20 years there will be a "brown" Christmas. At the end of February there is usually 22 cm of snow.

Rain occurs infrequently during the winter months. It accounts for about 5% of winter's total precipitation or 1% of the year's total. Occasionally, several millimetres will fall, and about once every 20 years there will be a single rainstorm during a winter month that will deposit up to 9 mm. When this happens in combination with freezing temperatures it creates

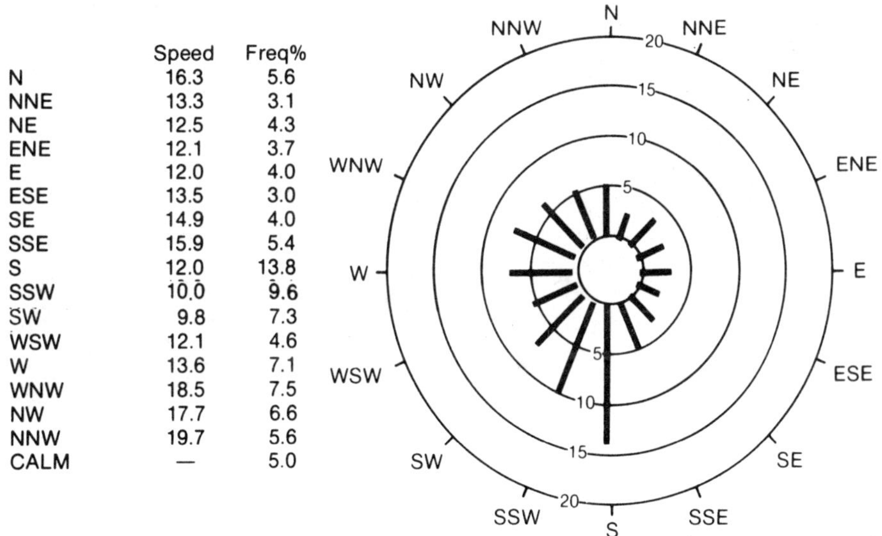

Figure 5. Winter winds

Downtown Edmonton, after snowfalls in 1903 and 1974 (E. Brown and Edmonton Journal Collections, the Provincial Archives of Alberta).

severe problems for motorists and pedestrians, as well as causing damage to power lines and trees.

Wind and Wind Chill

Winter winds average 13.2 km/h and they are usually the most gentle of the four seasons (Figure 5). The light winds are advantageous at this time of year because the discomfort caused by the combination of low temperatures and wind is reduced. This is commonly referred to as wind chill, which is an index of the rate that heat is lost from exposed flesh. The loss of body heat in cold weather increases with a rise in wind speeds. Thus, given a certain temperature a person will feel progressively colder as the wind increases, and prolonged exposure to windy conditions and low temperatures brings the risk of frostbite. Table 2 indicates the mean percentages of time that the wind chill exceeds various limits at Edmonton. A wind chill nomograph is provided as Appendix 3.

A combination of snow and blowing snow, brisk winds, and low temperatures may result in blizzard conditions. In Edmonton, blizzards occur very rarely and they do not usually last more than half a day. Outbreaks of cold arctic air associated with these storms result in maximum hourly wind speeds that

TABLE 2

Mean percentage of time that the wind chill exceeds various limits at Edmonton Municipal Airport

Class	Wind Chill Watt/m²	Oct.	Nov.	Dec.	Jan.	Feb.	Mar.	Apr.
Work and travel become uncomfortable unless properly clothed.	1000	25	70	92	94	85	73	34
Conditions no longer pleasant for outdoor activities on overcast days.	1200	3	37	66	78	58	46	9
Conditions no longer pleasant for outdoor activities on sunny days. Work and travel become more hazardous unless properly clothed. Heavy outer clothing necessary.	1400	*	18	42	62	30	22	2
Unprotected skin will freeze with direct exposure over prolonged period. Heavy outer clothing becomes mandatory.	1600	0	8	26	45	14	7	*
Unprotected skin can freeze in one minute with direct exposure. Multiple layers of clothing mandatory. Adequate face protection becomes important. Work and travel alone not advisable.	1900	0	*	6	13	1	*	0
Conditions for outdoor travel such as walking become dangerous. Exposed areas of the face freeze in less than 1 minute.	2300	0	0	1	*	0	0	0

* less than 1

exceed those of any other period of the year. The strongest wind gusts during these cold outbreaks are usually observed in January. Gusts of up to 114 km/h have been recorded.

Clouds

The variation of cloud cover throughout the year is depicted in Figure 6. Winter can be considered relatively cloudy compared to the other seasons, with each winter day averaging only 3.5 hours of bright sunshine. This is the least of the four seasons in terms of both actual hours and as a percentage of the theoretical maximum for the time of year.

Relative Humidity

The relative humidity expressed as a percentage is an index of the actual amount of moisture in the air compared to what the air could hold before dew or frost is deposited. Fully saturated air, such as occurs in clouds or fog, has a relative humidity of 100%, and the colder the air is the sooner it reaches this point. The relative humidity in winter averages 75%, which is the highest of the year. But, the actual amount of moisture in the air is still less than at any other time of year due to the low temperature of the air. As a result, the air inside a heated building will seem particularly

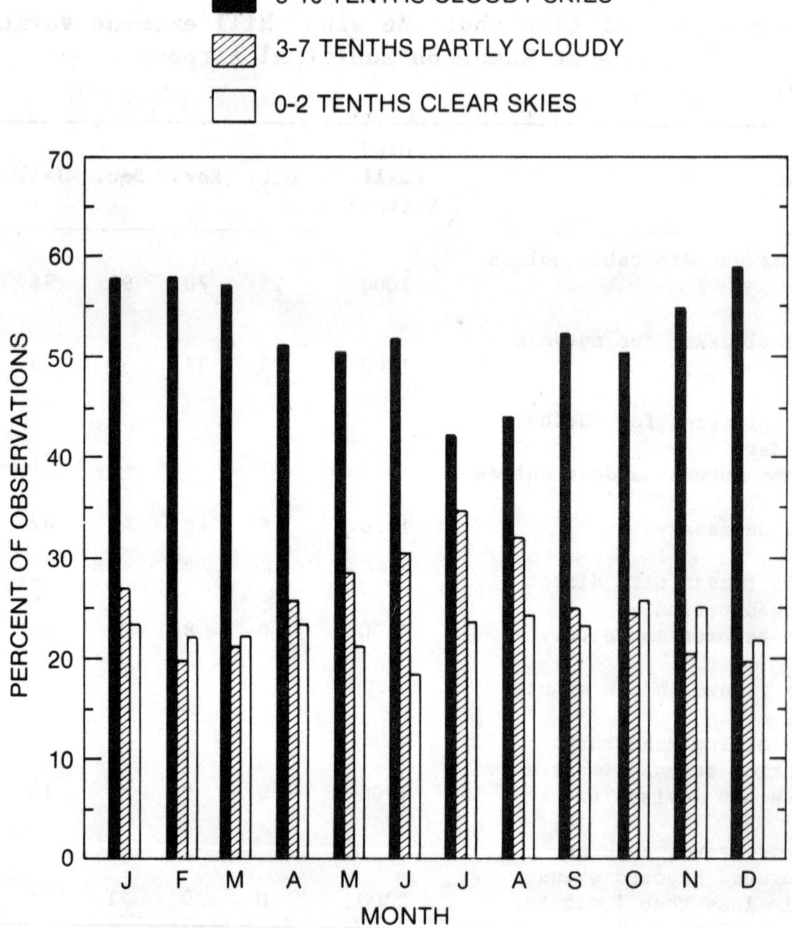

Figure 6. Per cent of average monthly cloud cover by class in Edmonton

dry unless water vapor is artificially added.

SPRING

The increase in the number of daylight hours in spring is accompanied by a rapid rise in temperature (Figure 7). In mid-March, the beginning of spring, the amount of daylight increases by about four minutes each day. Above freezing temperatures and relatively intense sunshine in early spring result in a rapidly disappearing snow pack. The mean snow depth at the end of March is 7 cm, but every second March averages less than a half centimetre on the ground. Exceptions to this occur, as in 1974, when there was 71 cm at month-end. Occasionally, there has been a trace of snow on the ground at the end of April.

Temperature

The mean temperature rises nearly 2°C each week during spring. It increases from -4.5° to 11.9°C as the season progresses (Figure 8). Spring temperatures exhibit a high degree of variability, especially at the beginning of the season. For instance, March 20 has had temperatures as low as -35.6°C and as high as 18.3°C, a range of 53.9°C. Readings from the end of spring are less variable; on 11 May

Fog in the North Saskatchewan River Valley. The Legislative Buildings are on the left and the central business district on the right (1976, Edmonton Journal Collection, Provincial Archives of Alberta).

1899 the record minimum of −4.4°C was set. When this temperature is compared with the 27.8°C record high set on the same day in 1949, it gives a range of 32.2°C. Thus, there is a difference of nearly 22°C in daily ranges of extreme temperatures between these two dates near the beginning and ending of the season.

The start of spring has been defined as the date that the average daily maximum rises above freezing. However, there can be spring days with the maximum temperatures less than 0°C. Occasionally, all the days in the last half of March will have afternoon temperatures below freezing. In April, there are usually 27 days with maximum temperatures above 0°C; however, in 1959 there were only 16 days. Since 1882 there has been a total of eight days in May for which the temperature did not reach 0°C. The last occurrence of this was in 1959, when the temperature remained below freezing for three consecutive days in early May of that year.

A cold period in spring can be similar to winter-like conditions. For example, a cool April might have a mean temperature approximating that of a mild February. In 1954 a cool spring was recorded with a mean daily minimum temperature of −5.4°C. Winter was still

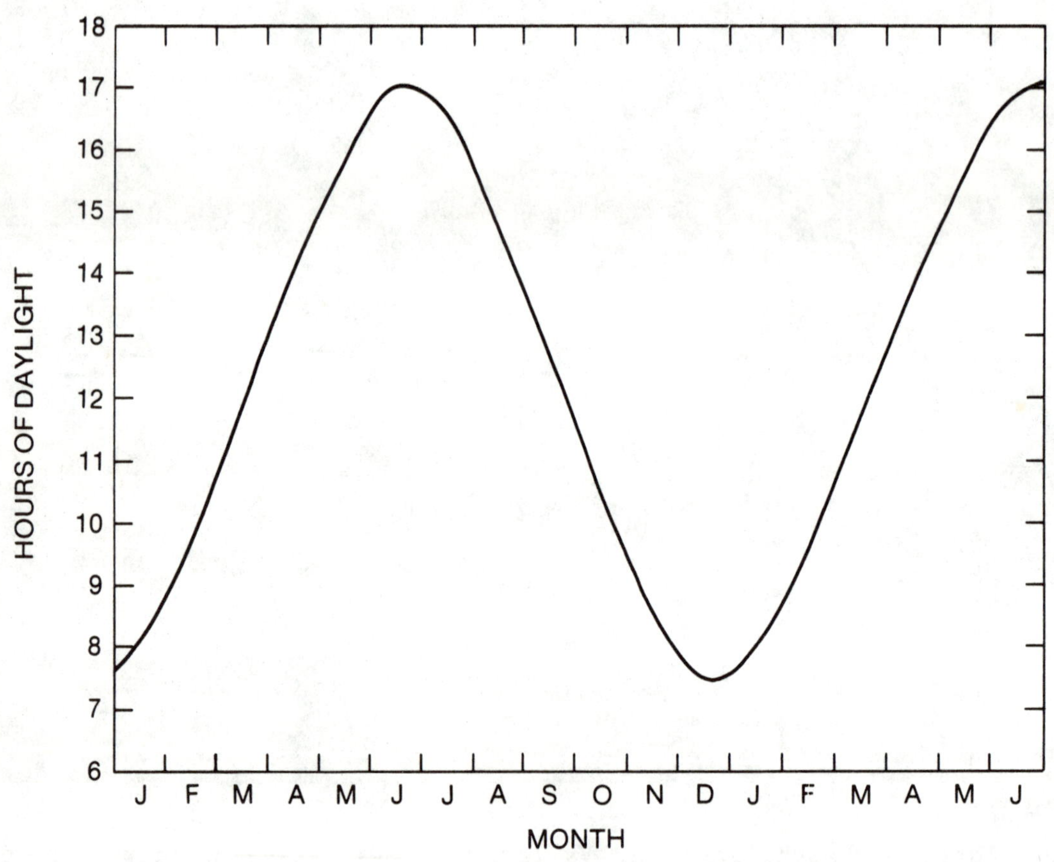

Figure 7. Hours of daylight throughout the year in Edmonton

in the air long after spring was supposed to have arrived.

The portion of spring that is in March will usually have frost on all but one night. An average April will have about 10 days with frost, but this figure has ranged from 9 to 30 days. On average, the first half of May will have two or three days with night-time temperatures at or below 0°C; one May in four will be frost free.

Each spring will have several days that are at or about 18°C during the warmest part of the day. March rarely has these summer-like conditions, but each April averages three days with values at or above 18°C. The number of mild days in April is variable and it is just as likely that there will be either one or five days with 18°C temperatures. The maximum number was 14, in 1915. Days with temperatures at or above 30°C are rare in spring. The earliest day in the year that 30°C has been attained was 25 April 1977. Two other days in April and four in the spring portion of May have had 30°C temperatures.

A warm month in spring can have higher mean daily maximum temperature than a normal summer month. For example, April's has been as high as the mean for June. One of the mildest springs was in 1980. The mean daily maximum was 15.4°C for the season. By

Early spring in Edmonton along the North Saskatchewan River (1981, Provincial Archives of Alberta).

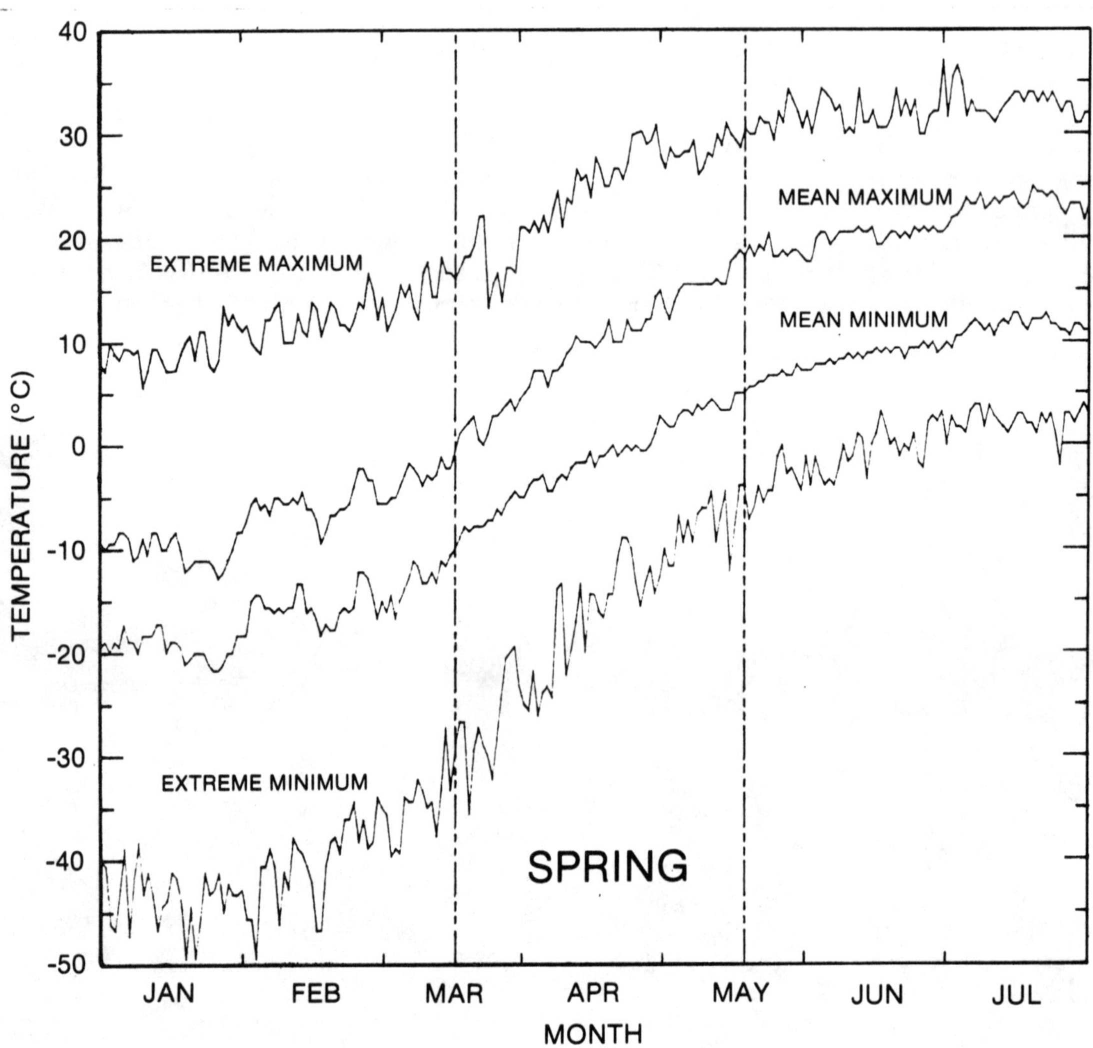

Figure 8. Mean daily and extreme temperatures during spring

the end of April in that year the vegetation was in full leaf and summer had
already arrived.

Precipitation

Spring is a dry period. It
accounts for 12% of the annual precipitation, with most of this falling in
the season's last three weeks. The
seasonal total is 55 mm, but the variability in this amount is much greater
than that of winter or summer. For
example, the normal precipitation in
April is 21.7 mm; however, one in ten
Aprils will either have less than 2 mm
or more than 71 mm. The driest April
was in 1882 with only 0.8 mm of rain.

The wettest was in 1953 with 88.1 mm.
This latter amount, however, was still
less than that for a typical July.

As spring advances there is a
shift in the type of precipitation
(Figure 9). At the beginning of the
season almost all the moisture falls as
snow; during April it is approximately
half snow and half rain; and by late
spring it is nearly all rain. Thunderstorms have never been recorded in
Edmonton during March, but they occur
once every 10 years in April.

There are usually seven days with
precipitation in April. Three of these
days will receive it in the form of

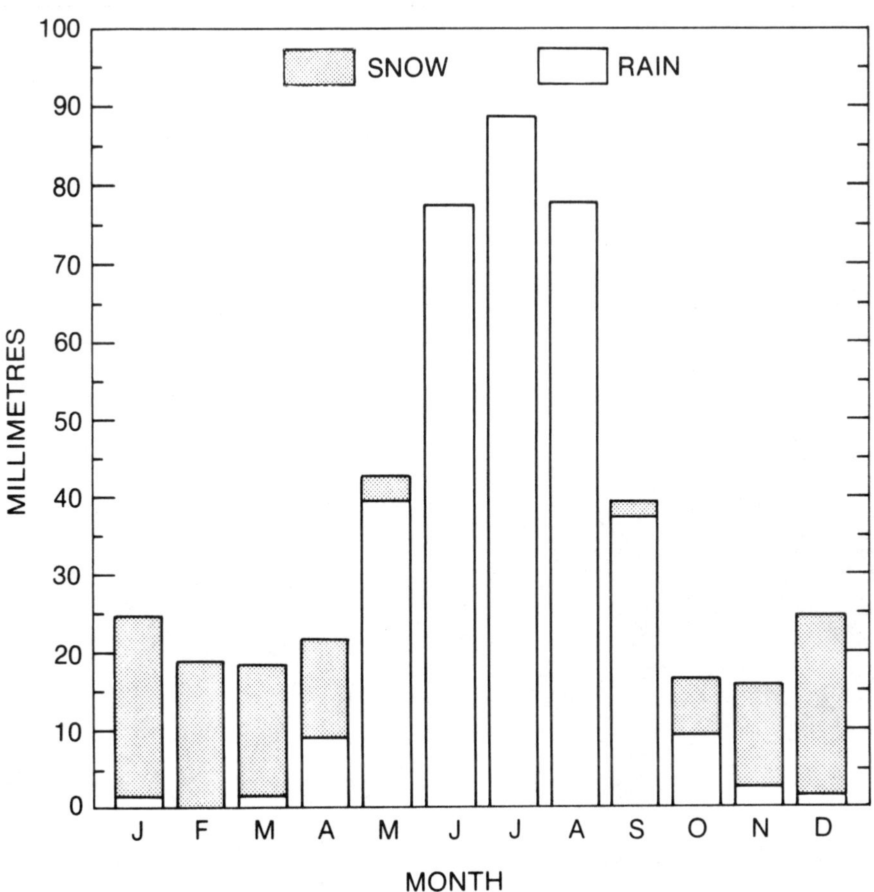

**Figure 9. Monthly distribution of snow and rain (Snow is expressed as water
equivalence where 10 mm of snow is approximately 1 mm of rain).**

snow, but it is as common to have four as it is to have no days with snow. An average snowstorm in April will deposit 4 cm; however, there is usually one that releases nearly twice this amount. The largest April snowstorm was in 1951, when 38.1 cm were dumped on the area. This was the second greatest snowfall ever recorded in Edmonton over a 24-hour period.

One day in April will usually have freezing precipitation in which a rain shower turns to snow as the event progresses. Rarely will a snow flurry turn into a rain shower. The remaining three precipitation events in a typical April will be in the form of rain. An April rain shower averages 3 mm, but again, one of these three storms will have about twice this amount. The extreme 24-hour April rain storm occurred in 1940 when 27.2 mm fell.

Wind

As spring advances mean wind speeds generally increase. They range from an average of less than 14 km/h in late March to more than 16 km/h in early May. Winds out of the northwest are generally the strongest, at about 21 km/h, while the southwest winds are the lightest, at 10 km/h. The winds during the entire season for all directions average 15.4 km/h (Figure 10). The strongest wind gusts in spring do not attain the velocities that they do

The effects of a mid-May snowfall in Edmonton. The Legislative Building is in the background (1944, A. Blyth, Provincial Archives of Alberta).

in the other seasons; however, the average winds speeds are the highest for the year. In part, this explains why the kite-flying season is at its peak at this time of year in Edmonton.

Clouds

In spring more than half the daylight hours have bright sunshine. The decrease in cloudiness in spring compared to winter allows 20% more of the potential bright sunshine to reach the surface. As the season progresses, daylight hours increase, which results in April and May averaging nearly eight and nine hours of bright sunshine per day, respectively.

Relative Humidity

Of the four seasons spring has the least amount of moisture in the air compared to what it could hold if it were saturated. The average relative humidity is 61% (Figure 11). The small

amount of atmospheric moisture and the mild daytime temperatures assist in the drying of the ground once the melt water from winter's snow has run off.

SUMMER

The start of summer corresponds to the gardening season in Edmonton. The long weekend in May, which falls between the 18th and 24th of the month, is recognized as being the optimum time for planting gardens because the frost hazard is minimal by this time of year.

Temperature

Summer temperatures have less variability than those at any other time of year. The mean seasonal value is 15.7°C (Figure 12). the warmest summer was in 1961 with a mean of 18.5°C. Daily maxima were about 25°C while minima were near 13°C. The hottest summer month on record was August 1981 with daily highs near 27°C

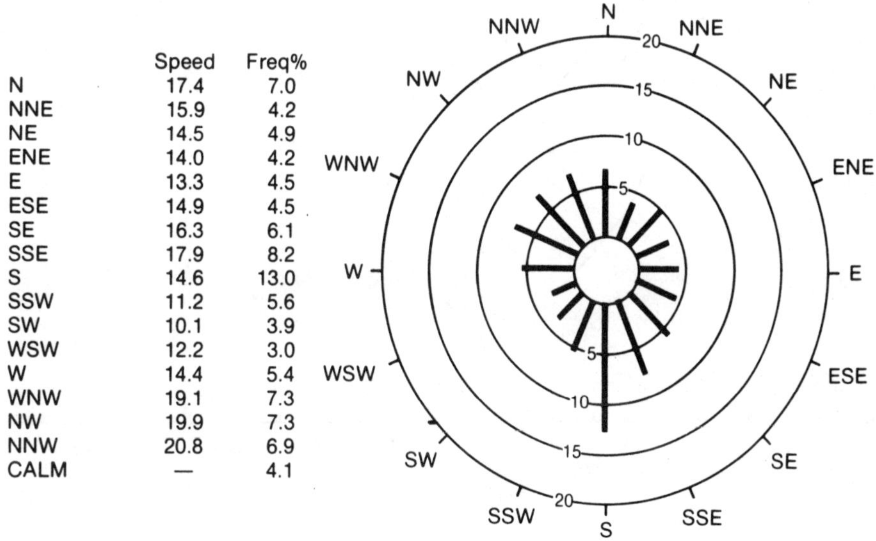

Figure 10. Spring winds

and lows near 14°C. The coolest summer was in 1887; the mean was 13.3°C with average highs of 20°C and lows of 6°C. In that year frost occurred three times in late May, five times in June, and once in August. The coolest summer month was June 1902, with daily highs of 18°C and lows of 5°C.

Rarely does any one month have all its daily maxima above 18°C. During July this happens about once every 10 years. June, July and August have had as few as 13, 22 and 16 days, respectively, with temperatures above 18°C. The averages are 22, 28 and 26 days. July is usually the warmest month of the year, and one July in two will have the year's highest temperature. Less than half the time June or August will have the year's highest, and once every five years it will occur in May or September. On average, the highest value of the year will fall between 31° and 32°C. It does, however, usually rise to 30°C or more three times per year, and there have been as many as 11 of these very warm days in a single year. About once every 20 years the temperature reaches 34°C. Since 1882, it has climbed to 35°C or more on four occasions. The highest value ever measured in the City was 37.3°C on 29 June 1937. The mean percentage of time that uncomfortably hot, humid conditions exist are listed in Table 3.

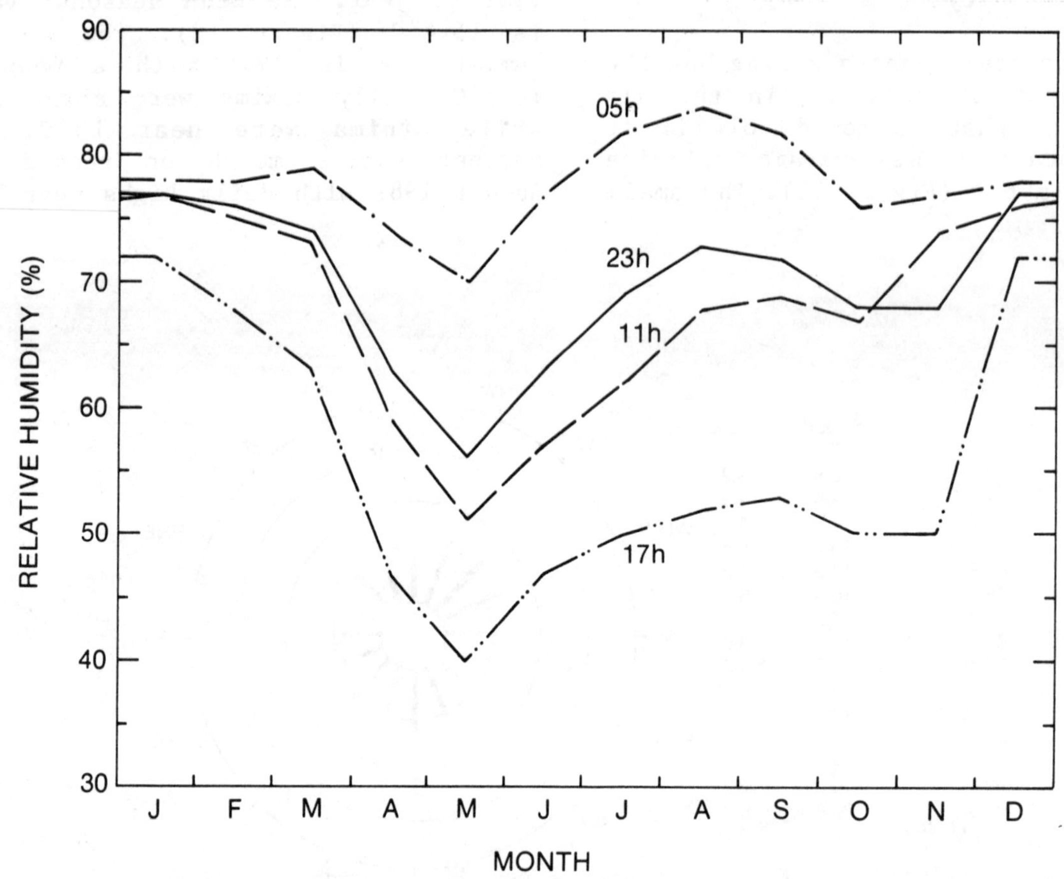

Figure 11. Average monthly relative humidities for selected hours of day

Frost

In Edmonton there are 180 days between the date that the mean daily minimum temperature rises to 0°C in the spring and the day before it drops below this value in the fall. But, the number of days separating the last spring and first fall frost is normally less than this due to the wide variability in minimum temperatures. Between 1951 and 1980 the frost-free season averaged 140 days. The 6th of May is the usual starting date of the frost-free period and September 24 is the average ending date. The earliest occurrence of the last frost in spring was April 10 and the latest was June 21. The earliest and latest dates of the first fall frost were September 6 and October 16, respectively. The longest interval with temperatures continuously above freezing was 184 days, from April 11 to October 11, 1980. The shortest was 42, from June 30 to August 13, 1884. The shortest frost-free period since the mid-1930s was 88 days in 1942.

Since 1881, 45 of the 111 summer dates have had extreme low temperatures below freezing and another nine at 0°C. The most recent dates with below freezing temperatures in early and late summer were 21 May 1959, when it was -0.6°C, and 6 September 1939, when it

Summer in Edmonton, looking east across the North Saskatchewan River and out over some of the western sections of the valley park system (1983, R. Olson).

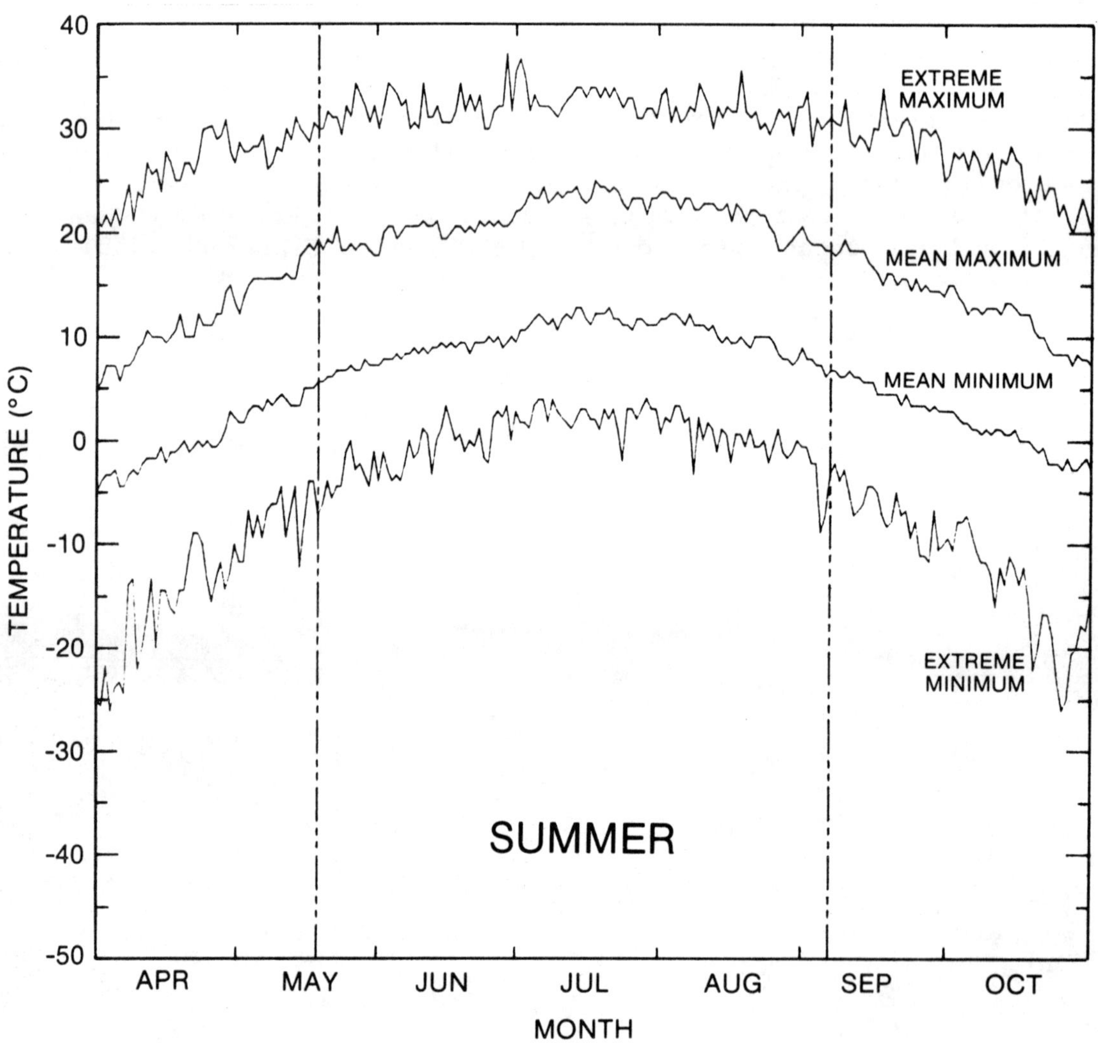

Figure 12. Mean daily and extreme temperatures during summer

dipped to -2.8°C. The last year frost was measured during any of the three complete summer months was June 1942. The 40 summers preceding that year averaged more than one frost during the June-to-August period alone, and there were only six summers that were frost free.

The main reason that summer frosts are now rare events in Edmonton is because of the moderating influence that the City has had on temperature. This is known as the urban heat island effect and it is the result of energy balance differences between the City and its surrounding area. The effect is particularly noticeable when the frost-free periods from Edmonton and Ranfurly, a town about 100 km to the east, are compared (Figure 13). The relatively long frost-free season in Edmonton can be almost entirely explained by the spread of the City since the discovery of oil near Edmonton in the 1940s.

Precipitation

Showers are common during the summer and account for nearly half of the year's total precipitation. The average for the season is 227 mm and most of it is received as rain, but snow has been recorded in all summer

TABLE 3

Mean percentage of time that uncomfortably hot, humid conditions exist during the summer at Edmonton Municipal Airport.

		Percentage of days with one hour or more with humidex greater than:		Percentage of hours with humidex greater than:	
	Period	30 (Some people are uncomfortable)	40 (Everyone is uncomfortable)	30 (Some people are uncomfortable)	40 (Everyone is uncomfortable)
May	21 - 31	0	0	0	0
June	1 - 10	2	0	*	0
	11 - 20	1	0	0	0
	21 - 30	2	0	*	0
July	1 - 10	6	0	1	0
	11 - 20	11	*	3	*
	21 - 31	6	0	1	0
Aug.	1 - 10	9	0	2	0
	11 - 20	5	0	1	0
	21 - 31	1	0	0	0
Sep.	1 - 10	*	0	0	0
	11 - 20	0	0	0	0

* less than 1

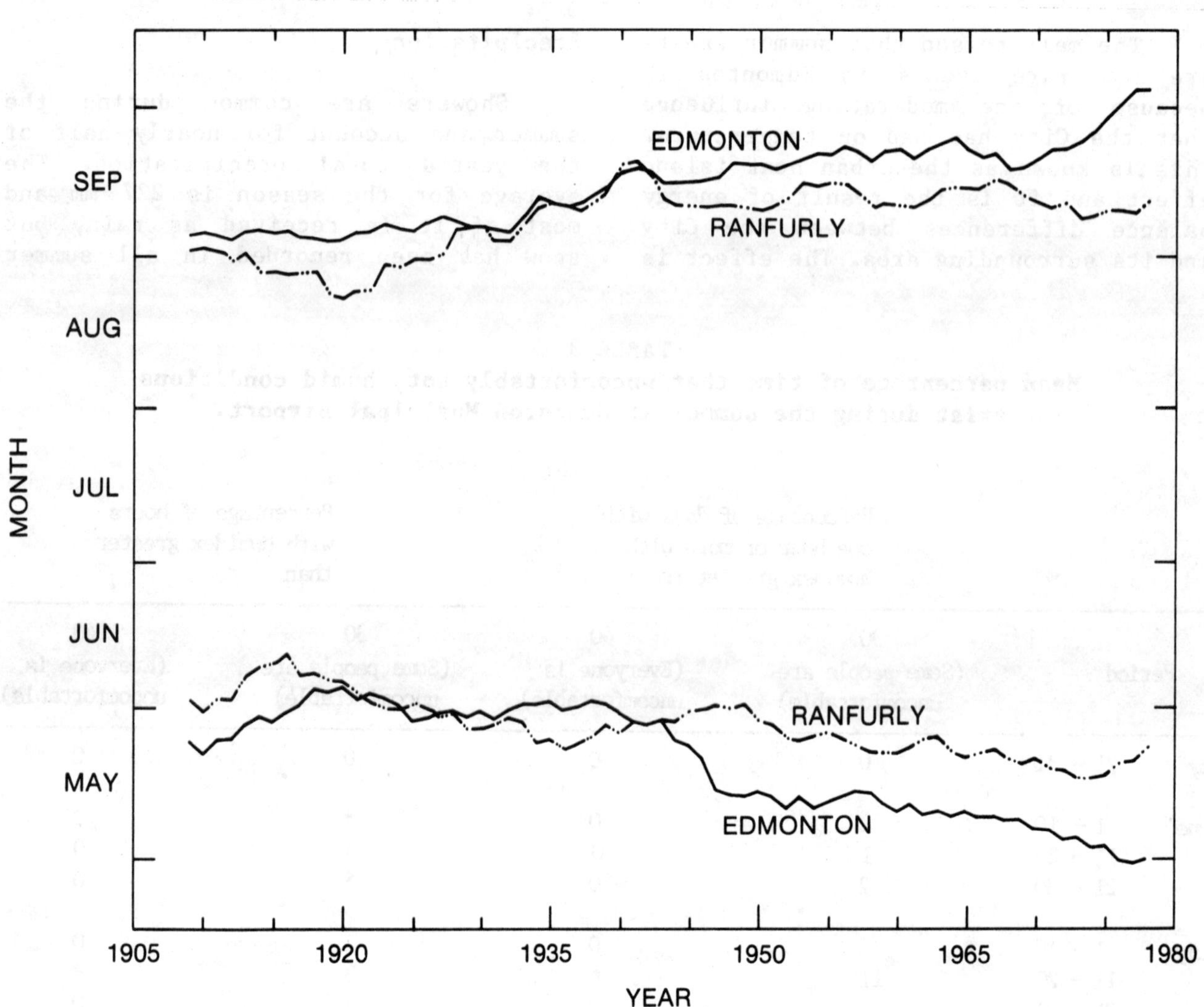

Figure 13. Frost-free periods at Edmonton and Ranfurly

months except July. Early and late summer storms are usually extensive in nature, while mid-summer rain events are often thunderstorms, which tend to be localized and have a relatively high rate of rainfall. The greatest single day rainfall occurred on the last day of July, 1953, when 114 mm fell. The extreme single day rainfalls for June and August are 80 and 60 mm, respectively.

Variability in the amount of precipitation can be expected from summer to summer. June 1981 was very dry with 13 mm of rain, a monthly total which occurs less than once every 20 years. July of that same year was very wet with 176 mm, again an amount anticipated only once every 20 years. The summer with the most precipitation was 1901, when more than 440 mm of rain fell. The driest was the summer of 1939, with a total of 111 mm. The average number of days with precipitation during summer is 48. Every second summer will have between 10 and 15 days each month with precipitation of 1 mm or more. The summer of 1886 had a total of only 10 days with rain. Occasionally, there may be a month in summer with measurable rainfall on two days out of three.

Thunderstorms, hail and tornadoes are generally limited to the summer months. On average there are 22 days each year with thunderstorms in the Edmonton area. Their frequency increases rapidly in the early part of the season, reaching a maximum of eight days in July. Many of the more severe thunderstorms are accompanied by hail. Hail is local in nature and often only a small part of the City will be affected, while the remainder of the City is untouched. On rare occasions tornadoes have been sighted, but extensive damage has never been reported from these events. By September, thunderstorm activity drops to one day per month, while hail and tornadoes are very infrequent.

Wind

Winds during summer are generally light, averaging 14.6 km/h (Figure 14). The strongest winds of the season are

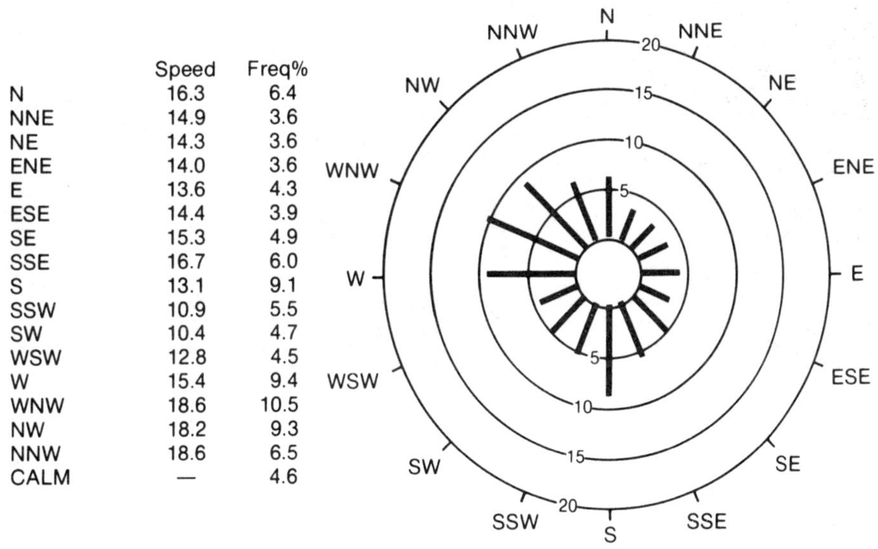

Figure 14. Summer winds

The June 1915 flood of the North Saskatchewan River. The Low Level Bridge (above) had to be weighed down with loaded gravel cars to prevent it from being washed away. The lower scene looks south across the flooded Rossdale district (top, City of Edmonton Archives; bottom, Provincial Archives of Alberta).

in June and the lighest are in August.
Winds from the west to northwest are
the most common and result in the
strongest single-hour winds of up to 65
km/h. The maximum gust speeds are in
June, and have been as high as 117
km/h.

Sunshine and Clouds

More than 1000 of Edmonton's 2264
average annual hours of bright sunshine
are received during the summer. The
mean for the season is approximately
nine hours per day; however, this value
is quite variable from summer to summer
due to the amount of cloud cover. Of
the individual summer months August is
the most variable. August 1959 was the
most cloudy of all summer months and it
had an average of 1.7 hours per day
with sunny skies. August 1955, at the
other extreme, had more than 10 hours
of bright sunshine each day.

Relative Humidity

The early morning hours in summer
have average relative humidities of
about 83%. This is higher than any
other time of day or year. In late
afternoon, the warmest part of the day,
this value drops on the average to
about 45%. Over the course of the summer, the average relative humidity is
quite low, yet the amount of moisture

Fall in Edmonton, looking south towards the Alberta Government Telephone Buildings
and the Hotel MacDonald (1973, Edmonton Journal Collection, Provincial Archives of
Alberta).

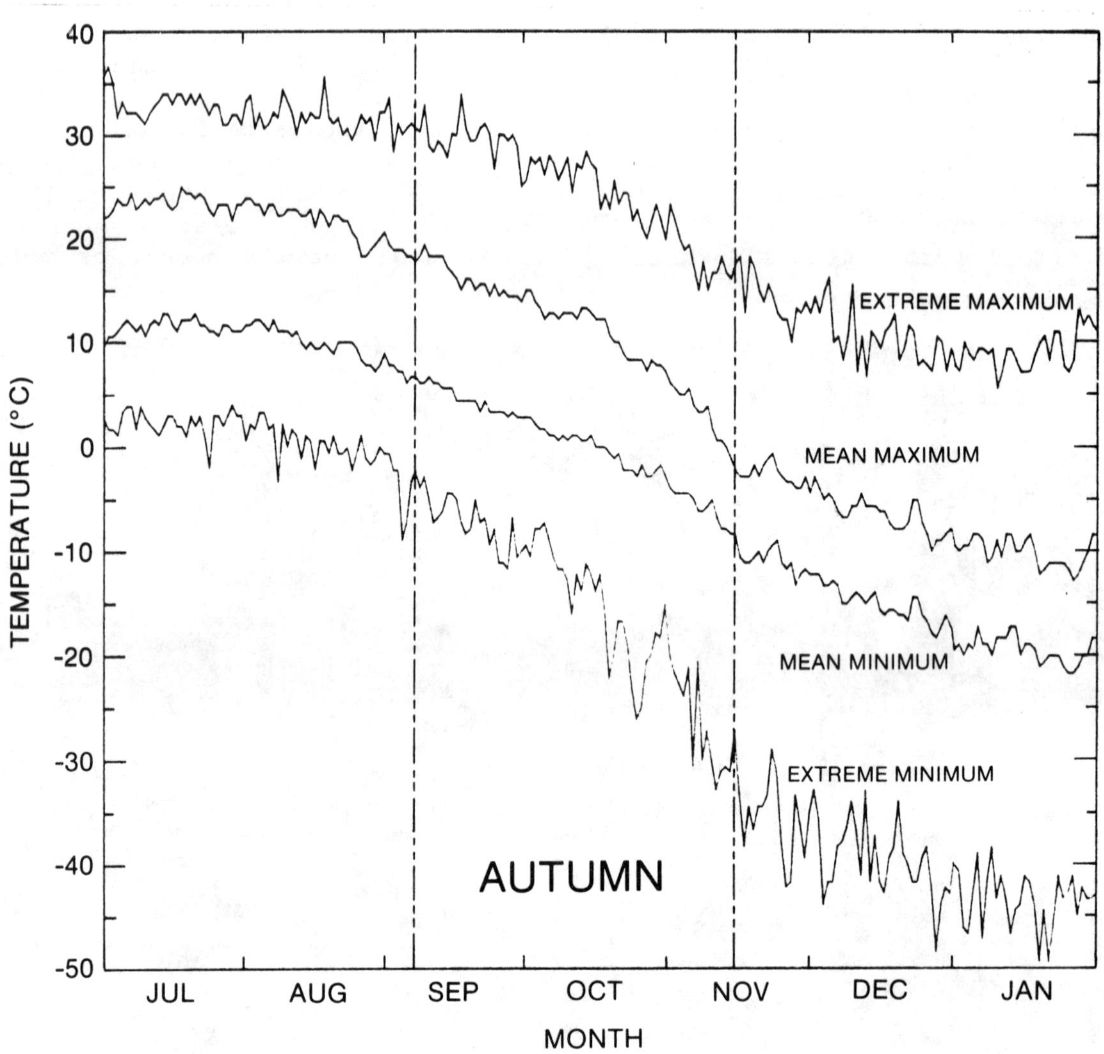

Figure 15. Mean daily and extreme temperatures during fall

in the air is the greatest at this time of year because of high temperatures and ample moisture supply.

FALL

Temperature

The beginning of fall, in early September, is generally summer-like. The end of fall, in mid-November, is usually not quite winter-like because of the lack of snow on the ground. The mean seasonal temperature is 5.9°C, with the average maximum and minimum 11.0° and 0.7°C, respectively. A characteristic of fall is the large variability in temperature. For example, the warmest fall was in 1963 with an averge afternoon temperature of 14.4°C. The coolest, in 1919, had an average daily minimum of -5.9°C. These values, however, are not that significant relative to the overall cooling that occurs as the season progresses (Figure 15).

There are always days in fall that are summer-like, with maximum temperatures above 18°C. The least number of these days was in 1941 when only four

were above 18°C. At the other extreme, half of the days in the fall of 1938 were 18°C or more. October usually has four days that are summer-like, but it is just as common to have none as it is to have 10. In November the temperature rarely gets above 18°C. There have been 19 such occasions since 1880 and two of those were in the winter portion of November. Prolonged mild spells usually occur every fall. These periods are often referred to as Indian summer and temperatures can be relatively high considering the time of year. September 1981 had two warm periods, each of which had three days in succession with maximum temperatures averaging above 30°C.

The mean daily minimum temperature drops to freezing on October 19, even though frost usually occurs several weeks prior to this date. September generally has three days with frost, October has 16, and all but two days in the fall portion of November typically have below freezing temperatures.

There are usually several days during fall in which the daily maximum

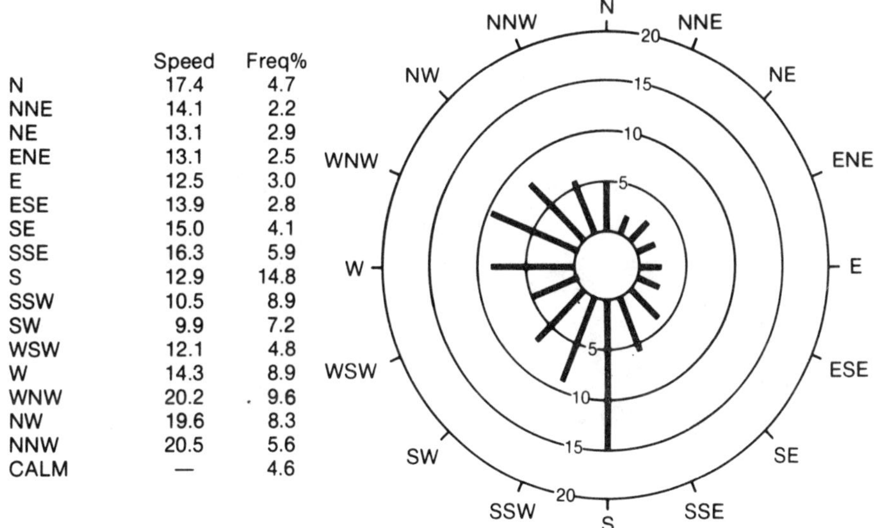

Figure 16. Fall winds

The North Saskatchewan River in late fall. Ice pans and fast ice are forming. The MacDonald Hotel and the Low Level Bridge are in the background (1940, A. Blyth Collection, Provincial Archives of Alberta).

does not reach 0°C. They are often accompanied by unseasonable snowfalls, especially if these cool fall days occur early in the season. Once every 10 years September will have a day that does not get above freezing. Three-quarters of all Octobers will have at least one day with winter-like temperature conditions. The average number is two. Nine-tenths of all fall portions in November will have at least one day with a maximum less than 0°C, while the average is four. Very cold weather can occur in fall. Towards the end of the season in 1940 there were three days in succession with lows of -30°C or less.

Precipitation

The monthly precipitation totals rapidly decrease during fall. The average for the entire season is 52.3 mm but more than half of this occurs in the first three weeks. The fall with the least precipitation was in 1929 with only 11.9 mm, while the greatest amount was 140.5 mm in 1978. Half of all Octobers have less than six days with precipitation. This is the least number of days of any month of the year and it promotes the continuation of summer outdoor sporting activities.

As fall progress the form of precipitation changes. It is almost all rain in September, but in November it is almost all snow (Figure 9). Usually one September in four will have a snowfall of at least 1 cm. In 1934 a September storm resulted in 22 cm of snow. On average, each October has one event that deposits up to 4 cm of snow, but on 25 October 1885 a snowstorm dumped just under a half metre of snow. This was Edmonton's greatest snowfall within a 24-hour period, and it was nearly twice the maximum amount ever observed during a single winter day snowstorm.

Freezing precipitation is rare in September. Since 1957 it has only occurred on one occasion during the month. It is more common in October, while November has had more freezing precipitation events than any other month of the year. Between 1957 and 1982 there have been seven occasions during October with freezing precipitation. Each November averages two days with freezing precipitation. November, 1962, had seven days with freezing precipitation, the most ever recorded in a single month in Edmonton.

Wind

Early fall has relatively high wind speeds (averaging 14.3 km/h) compared to the end of the season. The strongest winds generally blow from the northwest sector and average more than 20 km/h (Figure 16). The most persistent winds, on average, blow from the south.

Clouds

The changing characteristic of the weather patterns over central Alberta during autumn causes an increase in the amount of cloudy skies when compared to summer (Figure 6). The amount of daylight rapidly declines during this period too. At the beginning of the season there are 13.3 hours per day of daylight. Seventy days later, at the end of fall, there are 8.5 hours. October has the year's highest percentage of hours with clear skies. Half the time during this month there are clear to partly cloudy skies.

APPLIED CLIMATE

Climate affects life on a day-to-day basis. While man's adaptability is such that many changes in the weather are taken for granted, occasionally an unusual climatic situation plays havoc with the normal routine of life. Man adapts to the changing temperature regime by insulating himself with appropriate clothing and regulating the comfort of his shelter, whether home,

Late fall in downtown Edmonton. The clouds are low and there is a light covering of snow that has partially melted (1971, Edmonton Journal Collection, Provincial Archives of Alberta).

automobile or place of work. The transition of the seasons allows for adaptability that is accomplished with a minimum of strain. Unusual patterns of hot and cold, either of intensity or duration, however, cause discomfort and stress. Since the overall temperature patterns accumulated over a season are related to heating and air-conditioning use, statistics of this type are useful in determining average and extreme energy uses for these purposes. Recreation patterns are determined not only by temperature patterns, but also by patterns of other climate elements such as precipitation, cloud and wind. Climate affects gardening in obvious ways, as growth and plant development are directly related to the major climatic elements: temperature, precipitation and sunshine.

Comfort

Temperature is not the sole determinant of comfort, as wind, humidity and sunshine are also of major importance. In a climate such as Edmonton's, windy weather greatly increases the chill in winter when temperatures are low, while humid weather greatly increases summer discomfort when the temperature is high.

On about four days out of five in January, conditions are not pleasant for outdoor activities on overcast days unless a person is adequately clothed (Table 2). On almost half the days, however, unprotected skin will freeze with direct exposure over a prolonged period and heavy outer clothing is mandatory. Conditions are generally less severe in December and February. Variations in wind chill across the metropolitan area can be considerable. The greatest wind chill is usually in those suburban areas that are sparsely settled and exposed and least in the city centre. However, the funnelling effect of wind between large buildings can greatly increase the wind chill locally in urban areas. Also, normally windy areas would have higher average wind chill than normally sheltered areas.

Wind chill conditions at Edmonton are slightly more severe than at Calgary, where the frequency of chinooks ameliorates winter tempertures (see Appendix 7). Conditions at Edmonton, however, are significantly better than at cities farther to the east, such as Saskatoon, Regina and Winnipeg, where exposure to chilling winter winds is legend. Wind chill

TABLE 4
Mean heating degree-days below 18°C across the Edmonton area

	July	Aug.	Sep.	Oct.	Nov.	Dec.	Jan.	Feb.	March	April	May	June	Year	Sep.-May
Edmonton Int'l A	79	109	246	414	705	963	1068	829	765	443	246	124	5991	5679
Edmonton Municipal A	49	78	210	380	651	880	1022	780	714	413	209	99	5484	5258
Edmonton Namao A	60	91	228	401	687	922	1047	817	732	425	226	109	5738	5478
Edmonton Stoney Plain	70	101	238	397	666	990	1026	790	732	438	238	127	5713	5415
Edmonton Woodbend	88	113	258	422	690	927	1065	818	749	453	252	150	5982	5631
Ellerslie	75	105	238	417	686	934	1071	834	778	448	240	128	5953	5645

conditions at eastern Canadian cities are significantly better due to warming effects of the Great Lakes and the Atlantic Ocean. At Vancouver and Victoria, the Pacific Ocean acts as a strong ameliorating influence, and severe wind chill conditions are virtually unknown.

There are a number of different ways to measure summer discomfort. Humidex is a measure of discomfort that incorporates the temperature and a term that is proportional to the amount of water vapour in the air. In contrast to wind chill, however, it does not consider the effect of wind. The higher the humidex the more the discomfort. For a given temperature, the higher the humidity, the higher the added term, and therefore the higher the humidex. With a humidex of 30 some people are feeling discomfort, while by the time and the humidex reaches 40, most people are uncomfortable.

At Edmonton, heat combined with humidity is not the problem that it is at many eastern Canadian cities (see Appendix 8). Significant heat and humidity normally only occur from early July to mid-August, and even at the height of summer, generally only on about one day in ten does the humidex reach 30 for at least an hour (Table 3). Only the Atlantic and Pacific coast

cities are as comfortable as Edmonton in mid-summer.

Winter Heating

Most buildings and houses will not have to be heated if the mean daily temperature is greater than 18°C. A good indication of relative heating costs from one winter to another can be obtained by taking the departure of the mean daily temperature below 18°C on a day-by-day basis and accumulating these degree-day values through the winter. For the Edmonton area, winter heating degree-days below 18°C vary from about 5400 units at the centre of the city to 5900 units in suburban or surrounding areas remote from the heat-island effect of the urban areas (Table 4).

Heating degree-day values are roughly the same at Edmonton as at other Prairie cities, but significantly lower values occur over eastern Canadian cities, and particularly at Atlantic and Pacific coast cities (see Appendix 6).

Summer Cooling

Many buildings are air conditioned if the mean daily temperature is significantly greater than a threshold value, which depends, among other things, upon the insulation and method

TABLE 5
Mean cooling degree-days above 18°C across the Edmonton area

	May	June	July	Aug.	Sept.	Total
Edmonton International Airport	1	7	12	10	1	31
Edmonton Municipal Airport	3	12	32	21	2	70
Edmonton Namao Airport	2	9	24	16	1	52
Edmonton Stoney Plain	1	6	16	11	1	35
Ellerslie	1	6	13	9	1	30

of construction. While there is as yet no general consensus on what threshold value should be taken, an indication of relative cooling from one summer to another can be obtained by taking the departure of the mean daily temperature above 18°C on a day-by-day basis and accumulating these degree-day values through the summer. For the Edmonton areas, the summer cooling degree-days above 18°C vary from about 70 units at the centre of the city to as little as 30 units at suburban or surrounding areas (Table 5).

The number of cooling degree-days at Edmonton is hardly significant when it is realized that many eastern Canadian cities have a figure more than ten times greater (see Appendix 6). Calgary has roughly the same number as Edmonton, as do the Atlantic and Pacific coast stations.

Recreation

Virtually all types of outdoor recreation enjoyed by Canadians are practicable in Edmonton's climate. As

TABLE 6

Recreation season lengths at Edmonton Municipal Airport

SEASON OR PERIOD	DEFINITION	DATES
Spring Thaw	A two-week period following the termination of winter	April 19 to May 2
Complete Summer		
Spring Shoulder	Begins with termination of spring thaw Ends with start of high summer	May 3 to May 21
High Summer	Begins on date that the mean daily maximum temperature rises above 18°C Terminates on date that the mean daily maximum temperature falls below 18°C	May 22 to Sept. 6
Autumn Shoulder	Begins with termination of high summer Ends with onset of winter	Sept. 7 to Oct. 26
Winter	Begins on median date of first snow cover of 2.5 cm or more Terminates on median date of last snow cover of 2.5 cm or more	Oct. 27 to Apr. 18

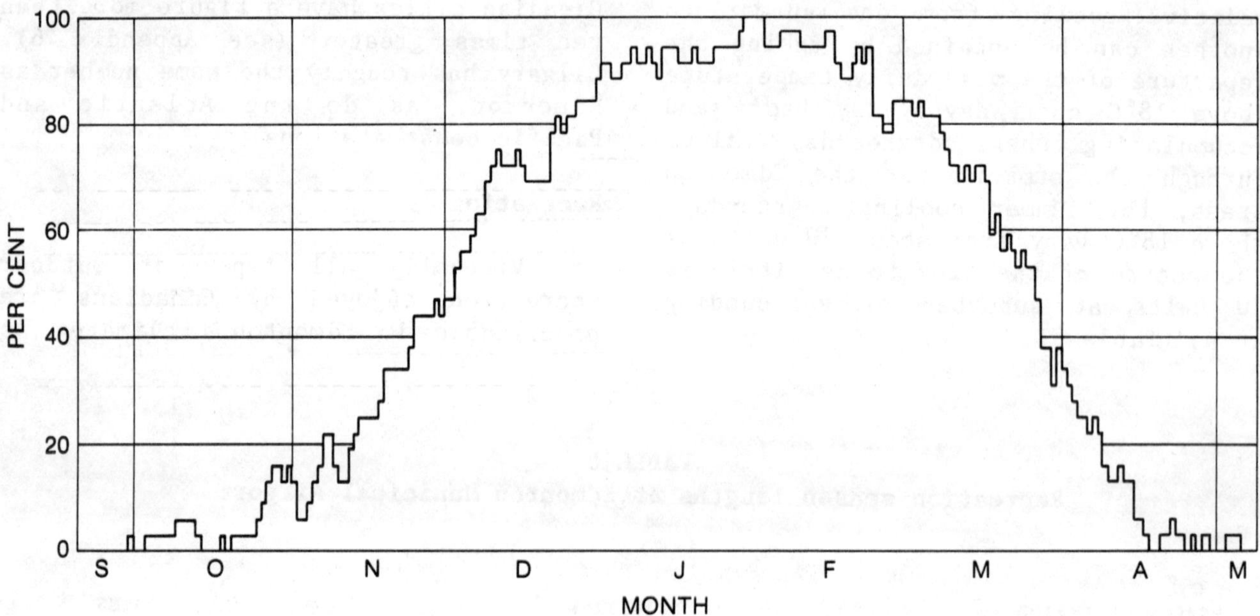

Figure 17. Percentage probability of a day with a suitable snow cover for winter recreation at Edmonton Municipal Airport

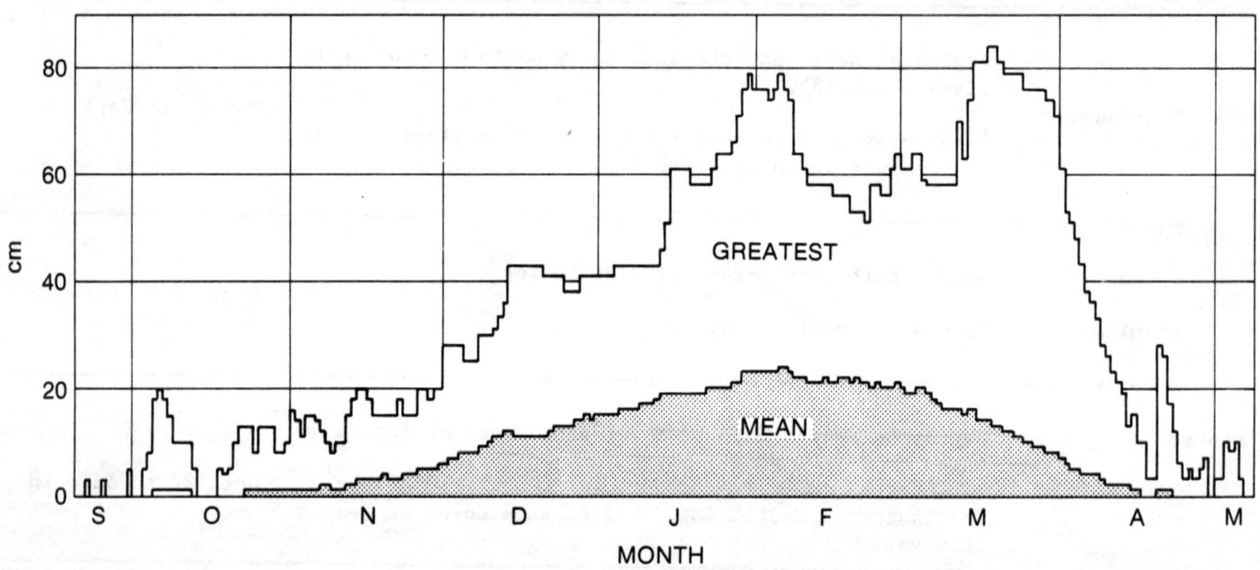

Figure 18. Mean and grestest daily depth of snow on the ground at Edmonton Municipal Airport

in all areas of Canada, these activities are highly seasonal. For general purposes, winter, spring, summer and fall seasons have been discussed earlier in this publication. For recreation and tourism specifically, a slightly different definition of winter and summer season (Table 6) is given below (Crowe, Baker and McKay, 1973).

The reader is referred to the publication, "The Tourism and Outdoor Recreation Climate of the Prairie Provinces" (Masterton, Crowe and Baker, 1976) for a full discourse on recreation seasons as well as on tourism and recreation climate in general in the Prairie Provinces.

Winter begins with the first snow, and this usually occurs about the end of October at Edmonton, although the first measurable snow cover has occurred as early as the last week of September or as late as mid-December (Table 7). Virtually all summer activities such as golf have ended, and the ground is too wet for much activity. Similarly, winter is defined to be over with the last snow cover, usually around mid-April, although the last measurable snow cover has occurred as early as the last week of February or as late as early May. During this winter period, the snow cover is unreliable until after the middle of November and after the end of March (Figure 17). For most of these four and one-half months, snowmobiling, skiing, and skating can be enjoyed provided the weather is suitable and the wind chill not too severe (Figure 18). Breaks in the snow cover are possible any time during the winter. However, they are rare after mid-December and before March. Snow cover data are highly site-specific. It can be assumed that the mean length of winter and the reliable snow-cover period would be somewhat shorter in the city core due to urban heating than it is at surburban and surrounding areas.

Following winter there is a period during which the ground is too wet for

TABLE 7
Snow cover season data for Edmonton Municipal Airport

	DATE		DATE		LENGTH (DAYS)	
	First 2.5 cm snow cover	First 2.5 cm snow cover lasting 7 days	Last 2.5 cm snow cover of continuous 7 days	Last 2.5 cm snow cover	First to last 2.5 cm snow cover (winter)	First to last 2.5 cm of 7-day periods (reliable snow cover)
Earliest (1952–1983)	Sep. 22/79	Oct. 3/57	Feb. 24/77	Feb. 24/77		
Shortest (1952–1983)					81(1976–1977)	81(1976–1977)
Mean (1955–1982)	Oct. 28	Nov. 17	Mar. 27	Apr. 14	169	131
Latest (1952–1983)	Dec. 11/81	Jan. 5/53	Apr. 20/67	May 6/59		
Longest (1952–1983)					215(1965–1966)	182(1957–1958)

much outdoor activity. Frost is coming out of the ground, and the sun is still not powerful enough to dry it out. This is the spring thaw period.

Many summer activities can be enjoyed from early May until the onset of winter at the end of October. The mid-portion of summer, when the mean daily maximum temperature is higher than 18°C, is essentially "shirt-sleeve" weather. This is called the "high summer". The complete summer then comprises a spring shoulder, high summer and autumn shoulder. In Edmonton, the high summer runs from about the last week of May to early September. Some summer activities are only practical in the high summer. These include swimming, beaching and sunbathing, when it is critical that temperatures be high. Other activities such as vigorous activities, including hiking and sports, and passive activities, including lounging and less active pursuits, can be enjoyed in the shoulder periods as well as during high summer.

Winter is roughly the same length at Calgary and other Prairie cities as it is at Edmonton (see Appendix 9), but it is significantly shorter at eastern Canadian cities and very much shorter at Vancouver and Victoria. The period of high summer is roughly the same length at Calgary, Vancouver and Victoria and at some cities along the Atlantic Coast as it is at Edmonton, but it is longer over the eastern Prairies and much longer over southern Ontario and southern Québec.

TABLE 8
Mean growing degree-days above 5°C across the Edmonton area

	Apr.	May	June	July	Aug.	Sept.	Oct.	Total
Edmonton International Airport	38	164	273	336	304	154	55	1324
Edmonton Municipal Airport	51	202	304	386	346	189	76	1554
Edmonton Namao Airport	43	186	290	368	328	172	62	1449
Edmonton Stoney Plain	42	175	273	350	312	167	68	1387
Edmonton Woodbend	37	157	251	331	289	150	57	1272
Ellerslie	39	172	270	341	306	160	56	1344

TABLE 9
Mean, earliest and latest dates of last killing frost in spring and first killing frost in autumn (-2°C) at Edmonton Municipal Airport

(1937-1972)	Earliest	Median	Latest
Last killing frost in spring	Apr. 3	May 3	May 29
First killing frost in autumn	Sept. 6	Sept. 28	Oct. 27

Gardening

In considering the climate with reference to gardening, the main elements, temperature, including frost-free period, sunshine, humidity and precipitation, are all of importance. While there are critical limits for these elements above which or below which growth is hampered, it is the effect of these elements operating together and timing which determines plant development and yield.

Most crops will not grow to any extent if the mean daily temperature is less than 5°C. A good indication of relative growth from one growing season to another can be obtained by taking the departure of the mean daily temperature above 5°C on a day-to-day basis and accumulating these degree-day values through the summer. For the Edmonton area, the growing degree-days above 5°C approximates 1500 units on the average (Table 8). The highest values occur in the city core and the lowest in suburban and surrounding areas. The growing season in the Edmonton area lasts from late April to early October, although growth is significant before mid-May and after mid-September.

Across Canada, significantly higher growing degree-day values in comparison with Edmonton occur in many regions (see Appendix 6), particularly over southern Ontario and southern Québec.

The practical growing season for the home gardener is determined by the occurrences of spring and autumn

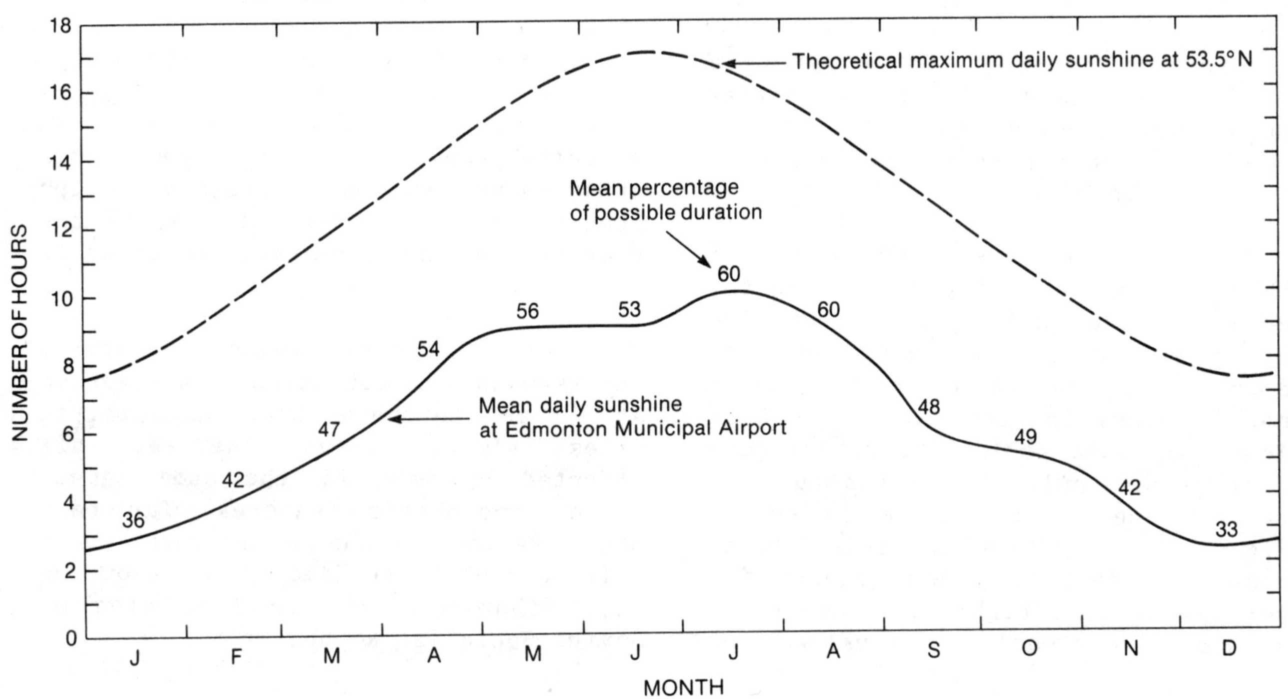

Figure 19. Mean daily sunshine at Edmonton and its theoretical maximum

frosts. On the average, the date of the last spring killing frost (-2°C) occurs in the first week of May for Edmonton Municipal Airport, but the earliest last killing frost occurred in early April, and the latest toward the end of May (Table 9). On the average, the date of the first autumn killing frost occurs in late September, but the earliest first killing frost occurred in early September and the latest toward the end of October. Ground frosts occur later in the spring and earlier in the autumn than killing frosts. Because of the variation in dates of last spring frosts from year to year, the home gardener would be well advised to avoid planting frost-susceptible varieties until at least the end of May. Frost data are very site specific. Low-lying areas are often more prone to frost, while locations on a side of a hill or near water bodies are less likely to receive frost. Late spring and early autumn frosts are also less likely in the urban core due to city warming than around the rural fringe.

Across Canada, the frost-free period (see Appendix 6) at many cities roughly approximates that for Edmonton. However, it is significantly longer at Vancouver and Victoria.

The duration of sunshine is an important factor in vegetative growth. The theoretical daily maximum sunshine is determined by latitude, and for Edmonton, it cannot be much greater than 17 hours in June. As a result of cloud, haze and pollution, actual sunshine occurs only slightly more than one-half the possible time in summer (Figure 19). In the months from June to August, critical to gardening, sunshine averages between eight and nine hours per day. Individual daily values, of course, range from no bright sunshine at all to close to the theoretical maximum. In these three months, Edmonton receives over 850 hours of bright sunshine in a normal year. Although this figure varies across the metropolitan area, the variation from place to place on the average is much less than the normal year-to-year variation.

Most cities in the Prairie Provinces have abundant sunshine, like Edmonton (see Appendix 10). Annual sunshine is much less at cities in eastern Canada and along the Pacific Coast.

Rainfall is necessary for crop growth, and adequate amounts usually occur in the summer months at Edmonton. In the case of unusually long dry spells, the home gardener can water his crops, but occasionally, long wet spells or unusually heavy rain will affect his crops adversely.

SIMILAR CLIMATES AROUND THE WORLD

There is a sense in which the climate of any sizable locality is unique. It is practical, however, to describe reasonably large geographic areas where the temperature and precipitation regimes throughout most of the year are roughly comparable to those of Edmonton.

The climate of Edmonton naturally corresponds to conditions experienced at other northern Canadian prairie sites, simply because they are all affected by many of the same large-scale atmospheric features. The only other region in the world that has a climate similar to Edmonton's is on the West Siberian Plain in the Union of Soviet Socialist Republics.

BIBLIOGRAPHY

CROWE, R.B., G.A. McKay and W.M. Baker, 1973. The tourist and outdoor recreation climate of Ontario, Volume One, objectives and definitions of seasons. Environment Canada, Atmospheric Environment Service, Publications in Applied Meteorology, REC-1-73, Toronto; 70 pp.

MASTERTON, J.M., R.B. Crowe and W.M. Baker, 1976. The tourism and outdoor recreation climate of the Prairie Provinces. Environment Canada, Atmospheric Environment Service, Publications in Applied Meteorology, REC-1-75, Toronto; 221 pp.

CLIMATIC DATA SOURCES

Environment Canada
Atmospheric Environment Service

Canadian Climate Normals 1951-1980
Vol. 1   Radiation, 1982
Vol. 2   Temperature, 1982
Vol. 3   Precipitation, 1982
Vol. 4   Degree Days, 1982
Vol. 5   Wind, 1982
Vol. 6   Frost, 1982
Vol. 7   Bright Sunshine, 1982
Vol. 8   Atmospheric Pressure, Temperature and Humidity, 1983
Vol. 9   Soil Temperature, Lake Evaporation, Days with Blowing Snow,
          Hail, Fog, Smoke/Haze, Frost, 1984

Climatic Normals 1951-1980
N.S. # 9-82 Edmonton Municipal Airport, 1982
N.S. #51-76 Edmonton International Airport

Principal Station Data
PSD-7   Edmonton International Airport, 1983
PSD-8   Edmonton Municipal Airport, 1983
PSD-9   Edmonton Namao Airport, 1983

Hourly Data Summaries
No. 23   Edmonton Namao Airport, 1967
No. 59   Edmonton Industrial (Municipal) Airport, 1968
No. 75   Edmonton International Airport, 1968

Daily Climatological Data
Edmonton (City and Municipal Airport), 1971

Daily Data Summaries
DDS-21   Edmonton (City and Municipal Airport) No. 21, 1972

Airport Handbook
Western Canada Vol. II, 1975

Freeze-up, Break-up and Ice Thickness in Canada
by W.T.R. Allen, CLI-1-77, 1977

Handbook on Agricultural and Forest Meteorology, R.A. Treidl, editor,
Vol. 1   1978
Vol. 2   1979
Vol. 3   1981

APPENDIX I
Suggested readings

Hare, F. Kenneth and Morley K. Thomas, 1974. Climate Canada. Wiley Publishers of Canada, Limited, Toronto; 256 pp.

Oke, T.R., 1978. Boundary layer climates. Methuen, London; 372 pp.

Longley, Richmond W., 1976. "Climate and weather patterns." W.G. Hardy (editor-in-chief). Alberta: a natural heritage. Published by The Patrons, distributed by M.G. Hurtig, Publishers, Edmonton; pp. 53-67.

Longley, Richmond W., 1972. The climate of the Prairie Provinces. Climatological Studies Number 13, Environment Canada, Atmospheric Environment Service, Toronto; 80 pp.

Longley, Richmond W., 1980. "The climate of Alberta." A Nature Guide to Alberta. Provincial Museum of Alberta Publication No. 5. Published by Hurtig Publishers in conjunction with Alberta Culture; pp. 26-28.

Schaefer, Vincent J. and John A. Day, 1981. A field guide to the atmosphere. The Peterson Field Guide Series. Houghton Mifflin Company, Boston; 259 pp.

APPENDIX 2
Glossary*

air mass – a widespread body of air that is approximately homogeneous in its horizontal extent, particularly with reference to temperature and moisture distribution. Air masses are classifed according to where they originated.

blizzard – a severe weather condition characterized by low temperatures and strong winds bearing a great amount of snow which is often picked up from the ground.

bright sunshine – a measurement of intense sunshine using a Campbell-Stokes sunshine recorder. The principle of this recorder is similar to burning a sheet of paper with a magnifying glass. The daily period of bright sunshine is less than that of visible sunshine because the sun's rays are not intense enough to register on the recorder just after sunrise, near sunset, and under cloudy conditions.

chinook – produced when the atmospheric circulation is sufficiently strong to force air from the west over the Rocky Mountains. The warmth and dryness is due to heating by compression as the air descends the mountain slopes (similar to the way a bicycle pump heats up).

climate – the long-term manifestations of weather – the synthesis of day-to-day weather variations in a locality. The climate of a specified area is represented by the statistical collective of its weather conditions during a specified interval of time, which is usually taken to include the following weather elements: temperature, precipitation, humidity, sunshine, and wind velocity.

climatic controls – relatively permanent factors which govern the general nature of the climate in a locality. They include: (a) solar radiation, especially as it varies with latitude; (b) distribution of land and water masses; (c) elevation and large-scale topography; (d) ocean currents; and (e) general atmospheric circulation.

continental climate – the climate that is characteristic of the interior of a land mass of continental size. It is marked by large annual, daily, and day-to-day ranges of temperature, low relative humidity, and by a small and irregular precipitation regime. Where a mountain barrier lies across the prevailing winds, as in Alberta, the effects of continentality are magnified.

---

*Adapted from <u>Glossary of Meteorology</u> (R.E. Huschke, American Meteorological Society, 1970) and <u>A Dictionary of Geography</u>, (W.G. Moore, Penguin Books, 1974).

frost - in Canada frost is said to occur when the air temperature is at or below the freezing point of water. The term frost is also used to describe the icy deposits which may form on the ground or on other objects such as car windshields.

frost-free season - the period between the last spring frost and the first fall frost.

front - a transition zone on the earth's surface between cold and warm air masses. Generally, a front is produced by the horizontal movement of two such air mass, which have originated from widely separated regions. The warm air, being lighter than cold air, is continually ascending this frontal surface.

freezing precipitation - supercooled water drops of drizzle, fog, or rain, which freeze on impact with the ground or other objects.

growing season - the part of the year when the growth of the natural vegetation is made possible by the favourable combination of temperature and precipitation.

hail - solid precipitation consisting of ice pellets that have a diameter of at least 5 mm. Hail is caused by the rapid ascent of moist air; the water drops freeze, and the size of the pellets increases as more water vapour freezes onto their surfaces. When the hail pellets are heavy enough to overcome the resistance of the ascending air currents, they fall, often growing by accumulating supercooled water droplets in the moist air.

humidity - the state of the air with respect to the water vapour contained in it (see relative humidity).

Indian summer - a warm, calm spell of weather, occurring in fall, especially in October. Such warm spells do not occur every year but when they do, the weather is reminiscent of the balmier days of summer.

precipitation - in climatology, the deposits of water, in either liquid or solid form, that reach the earth from the atmosphere.

relative humidity - the ratio between the actual amount of water vapour in the air and the amount which would be present if the air was saturated at the same temperature.

season - a period of the year which is characterized by special climate conditions. Seasons are caused by the inclination of the earth's axis and the revolution of the earth about the sun. In high latitudes there are four seasons: spring, summer, fall, and winter. However, the transition between winter to summer, and vice-versa, is so sudden that spring and fall largely disappear. In low latitudes temperature is not used to differentiate seasons; precipitation is used to divide the year.

solstice – the time during winter or summer when the noon-day sun is vertical above the point which represents its farthest distance north or south of the equator. The days are longest and the nights are shortest in the northern hemisphere on the summer solstice, while the opposite is true for the winter solstice.

weather – a condition of the atmosphere at a certain time or over a short period, as described by various meteorological phenomena.

wind chill – the combination of low temperatures with strong winds creating severe cooling stress. The term is used to describe the cooling effect of the wind at various temperatures.

APPENDIX 3
Wind-chill Nomograph

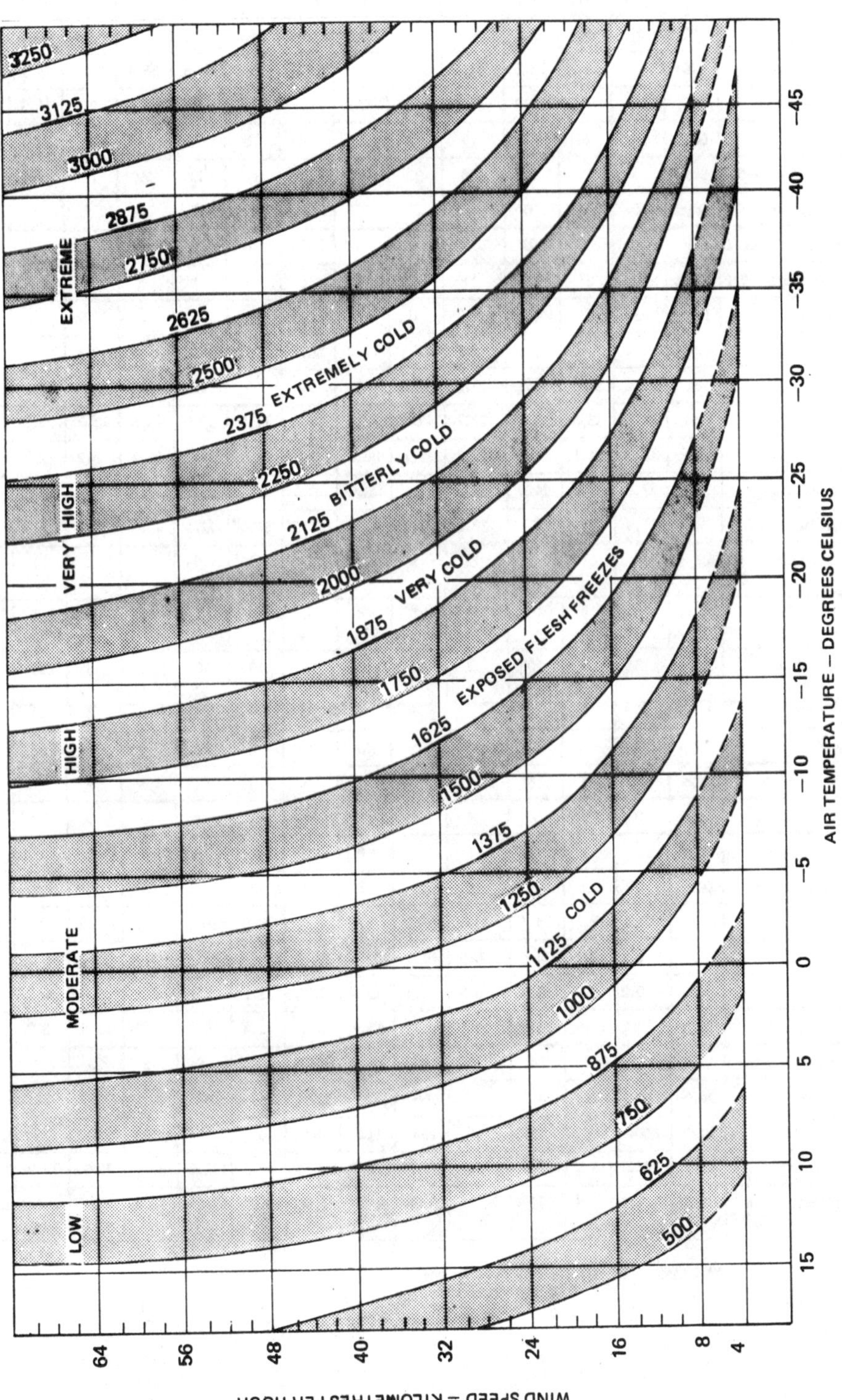

To determine the wind-chill factor follow the temperature up and the wind speed across until the two lines intersect. The value of the wind-chill factor can be interpolated using the labeled wind-chill factor curves. For example, at –10°C with a wind speed of 32 km/h the point of intersection lies between 1500 and 1625, or approximately 1570. It is not recommended that wind-chill factors be calculated for wind speeds below 8 km/h, since it is difficult to determine wind-chill factors at these wind speeds and because other factors such as relative humidity become important.

APPENDIX 4
Climatic normals for Edmonton Municipal A

Period	Element	Unit	Jan.	Feb.	Mar.	April	May	June	July	Aug.	Sep.	Oct.	Nov.	Dec.	Annual
1951–1980	Mean maximum temperature	°C	-10.7	-4.8	-0.3	9.6	17.2	20.7	23.0	21.6	16.5	11.1	0.4	-6.2	
1951–1980	Mean minimum temperature	°C	-19.2	-14.3	-9.7	-1.2	5.4	9.5	11.8	10.7	5.6	0.4	-7.8	-14.5	
1951–1980	Mean from maximum/minimum	°C	-15.0	-9.6	-5.0	4.2	11.3	15.1	17.4	16.2	11.0	5.8	-3.7	-10.4	3.1
**1880–1984	Extreme maximum temperature	°C	13.9	16.7	22.2	31.1	34.4	37.2	36.7	35.6	33.9	28.3	23.3	16.1	37.2
**1880–1984	Extreme minimum temperature	°C	-49.4	-49.4	-40.0	-26.1	-12.2	-3.9	-1.7	-3.3	-11.7	-26.1	-42.2	-48.3	-49.4
1951–1980	No. of days with frost	Days	31	28	29	19	2	0	0	0	3	14	28	31	185
1951–1980	No. days maximum temperature > 0°C	Days	6	10	17	28	31	30	31	31	30	29	17	9	269
1951–1980	No. days maximum temperature >18°C	Days	0	0	0	3	14	21	27	24	12	5	*	0	106
1951–1980	No. days maximum temperature >30°C	Days	0	0	0	*	*	1	1	1	*	0	0	0	3
1951–1980	No. days minimum temperature <-2°C	Days	30	27	26	11	1	0	0	0	1	8	24	30	158
1951–1980	No. days minimum temp. <-20°C	Days	15	7	3	*	0	0	0	0	0	0	2	8	35
1951–1980	Mean rainfall	mm	1.4	0.5	1.6	8.8	39.4	77.3	88.7	77.9	37.0	9.1	2.5	1.4	345.6
1951–1980	Mean snowfall	cm	27.2	21.3	18.7	13.2	3.1	T	0.0	T	2.2	7.5	15.4	27.1	135.7
1951–1980	Mean total precipitation	mm	24.6	18.8	18.5	21.7	42.5	77.3	88.7	77.9	39.1	16.6	15.7	24.7	466.1
**1880–1984	Greatest rainfall in 24 hrs.	mm	16.0	4.1	7.6	36.3	71.1	80.0	114.0	68.1	52.1	26.2	8.9	12.7	114.0
**1880–1984	Greatest snowfall in 24 hrs.	cm	27.9	27.9	24.1	38.1	38.1	3.0	0.0	2.5	22.1	40.6	39.9	25.8	40.6
**1880–1984	Greatest precipitaton in 24 hrs.	mm	27.9	27.9	24.1	38.1	71.1	80.0	114.0	68.1	52.1	40.6	39.9	22.9	114.0
1951–1980	No. of days with measurable rainfall	Days	1	1	1	3	9	13	13	13	9	4	2	1	70
1951–1980	No. of days with measurable snowfall	Days	13	9	10	5	1	*	0	0	1	2	7	11	59
1951–1980	No. of days with measurable precipitation	Days	13	9	10	7	10	13	13	13	10	6	8	12	124
1951–1980	No. of days with snowfall ≥ 1 cm	Days	8	6	6	3	1	*	0	0	1	2	4	7	23
1951–1980	No. of days with snowfall ≥ 10 cm	Days	*	*	*	*	*	0	0	0	0	*	*	1	1
1951–1980	No. of days with prec. > 10 mm	Days	*	*	*	*	1	2	3	2	1	*	*	*	9
1951–1980	No. of days with prec. > 25 mm	Days	0	0	0	*	*	1	1	1	*	0	0	0	3
1951–1980	No. of days with freezing prec.	Days	1	1	1	1	*	*	0	0	*	*	2	2	8
1951–1980	No. of days with a thunderstorm	Days	0	0	0	*	2	5	8	6	1	*	0	0	22
1951–1980	Degree-days below 18°C		1021.5	779.7	713.7	413.4	209.1	98.7	48.7	78.0	210.4	379.9	650.6	880.0	5483.7
1951–1980	Degree-days above 5°C		0.1	0.4	2.6	50.5	201.9	303.6	385.7	346.1	189.2	75.9	4.2	0.2	1560.4
1951–1980	Degree-days above 18°C		0.0	0.0	0.0	0.3	2.8	12.2	31.5	21.1	1.7	0.1	0.0	0.0	69.7
1955–1980	Average wind speed	km/h	13.0	13.0	13.8	15.5	16.1	15.4	14.0	13.4	14.6	14.3	13.5	13.0	14.1
1955–1980	Prevailing wind direction		S	S	S	S	S	WNW	WNW	S	S	S	S	S	S
1955–1980	Maximum hourly wind speed	km/h	68	71	61	61	68	69	60	69	68	64	64	64	71
1955–1980	Maximum gust speed	km/h	114	105	93	90	103	117	108	109	117	113	100	98	117
1951–1980	Bright sunshine	Hours	90.0	116.3	167.5	228.3	277.6	271.7	306.4	276.8	182.2	161.8	107.2	77.9	2263.7
1967–1976	Mean daily global radiation on a horizontal surface (calculated)***	MJ m⁻²	3.922	7.278	12.234	17.251	20.684	21.433	21.967	18.796	12.886	8.087	4.252	2.821	12.654

* less than 1 day	*** Edmonton Int'l A	> greater than
** Extreme values for earlier Edmonton stations (1880–1937) combined with Municipal A records (1937–1984)	T trace	≥ greater or equal than ≤ less than

APPENDIX 5
Climate data for Edmonton and surrounding areas

	Length of Record (yrs.)	TEMPERATURE (°C)					Rain-fall (mm)	Snow-fall (mm)	Total (mm)	Number of days with:				Number of days with fog	Frost Free period (Days)	DEGREE DAY (°C)							Bright Sun-shine (hours)	WIND		
		Mean Max	Mean Min	Mean	Extreme Max	Extreme Min				Rain	Snow	Precip	Thunder Storms			Above 24.0	Above 18.0	Above 10.0	Above 5.0	Above 0.0	Below 0.0	Below 18.0		% Calm	Most Frequent Direc.	Mean Speed in all Direc. (km/h)
Calmar	65	8.5	-4.1	2.2	36.7	-49.4	363.2	125.5	491.5	64	45	109	M	M	112	0.0	32.5	615.3	1388.0	2397.8	1566.4	5776.4	M	M	M	M
Edmonton International A	21	7.8	-4.6	1.6	35.0	-48.3	352.9	137.9	466.6	74	61	123	27	20	105	0.1	30.6	575.7	1328.0	2307.1	1692.7	5990.6	2314.9	7.7	S	13.4
Edmonton Municipal	44	8.2	-1.9	3.1	34.4	-48.3	345.6	135.7	466.1	70	59	124	21	16	140	1.0	69.7	754.3	1560.4	2600.7	1440.2	5483.7	2263.9	4.7	S	14.1
Edmonton Namao A	26	7.5	-2.8	2.4	33.9	-42.2	325.7	131.6	450.9	71	57	124	22	25	129	0.4	52.2	677.8	1453.3	2461.1	1572.2	5738.2	2280.4	6.0	W	16.7
Edmonton Stony Plain	15	7.4	-2.6	2.4	33.9	-40.0	391.4	131.1	528.8	74	57	130	20	M	137	0.0	35.6	621.8	1395.1	2407.7	1505.6	5712.6	M	0.4	NW	11.3
Edmonton Woodbend	7	8.1	-4.7	1.7	32.2	-46.0	388.3	135.1	527.2	68	42	111	M	M	87	0.5	M	562.0	1278.0	2274.4	1640.4	5981.8	M	M	M	M
Ellerslie	17	7.6	-4.1	1.7	33.8	-47.8	338.6	113.6	452.2	72	54	122	M	M	109	0.1	30.0	587.0	1347.8	2389.9	1679.2	5953.3	M	0.5	NW	12.7
Fort Saskatchewan	22	8.1	-3.5	2.3	35.6	-45.6	325.8	99.7	422.7	66	30	94	M	M	109	0.7	52.2	684.1	1470.1	2477.2	1635.2	5783.0	M	M	M	M

M = missing

APPENDIX 6
Canadian comparative temperature, precipitation, degree-day and frost data

	TEMPERATURE (°C)					TOTAL PRECIPITATION (mm)			WINTER SNOWFALL (cm)	MEAN ANNUAL degree/days			FROST		
	January Mean daily Max.	January Mean daily Min.	Annual absolute minimum	July Mean daily Max.	July Mean daily Min.	Annual absolute maximum	Jan.	July	Year		heating (<18°C)	cooling (>18°C)	growing (>5°C)	mean dates last spring	mean dates first autumn	mean frost free period days
EDMONTON																
Municipal A	-10.7	-19.2	-49.4	23.0	11.8	37.2	24.6	88.7	466.1	135.7	5484	70	1560	May 6	Sep. 24	140
International A	-10.9	-22.0	-48.3	22.4	9.2	35.0	24.4	91.6	466.6	137.9	5991	31	1328	May 25	Sep. 8	105
Namao A	-11.2	-20.0	-42.2	22.6	11.1	33.9	24.8	76.3	450.9	131.6	5738	52	1453	May 13	Sep. 20	129
Calgary	-6.0	-17.6	-45.0	23.3	9.4	36.1	16.2	65.4	423.8	152.5	5365	38	1387	May 25	Sep. 15	112
Cold Lake	-14.0	-23.9	-48.3	22.9	10.9	36.1	22.1	85.6	459.9	135.4	6166	46	1397	May 23	Sep. 6	105
Coronation	-11.6	-21.3	-44.4	24.0	10.5	37.8	21.5	63.0	373.6	137.0	5879	65	1459	May 21	Sep. 14	115
Fort Chipewyan	-20.3	-31.8	-50.0	22.4	9.7	34.4	19.1	75.6	380.7	158.4	7600	39	1159	June 1	Sep. 1	91
Fort McMurray	-16.5	-27.1	-50.6	23.1	9.5	36.1	22.7	75.4	471.9	163.8	6661	36	1290	June 7	Aug. 31	84
High Level	-18.5	-30.6	-50.6	22.8	8.7	34.4	20.6	68.9	386.7	163.6	7296	18	1176	June 14	Aug. 25	71
Lethbridge	-4.5	-16.0	-42.8	26.1	11.0	39.4	23.6	43.5	422.7	175.8	4745	116	1776	May 17	Sep. 19	124
Medicine Hat	-7.0	-18.3	-46.1	27.3	12.4	42.2	22.7	40.4	347.9	125.5	4868	187	1943	May 15	Sep. 22	129
Peace River	-14.9	-25.6	-49.4	22.3	9.0	36.7	22.1	60.5	375.1	146.4	6469	21	1239	May 31	Sep. 2	93
Red Deer	-9.9	-21.0	-50.6	23.0	9.1	37.2	23.8	77.7	463.5	138.9	5870	30	1323	May 26	Sep. 10	106
Rocky Mtn. House	-7.4	-18.5	-43.9	22.0	8.6	33.3	27.6	93.1	476.0	187.8	5613	19	1245	May 29	Sep. 7	100
Vermilion	-13.1	-23.6	-50.6	23.2	9.8	37.2	18.7	75.1	415.0	105.8	6168	38	1371	June 1	Sep. 9	100
Wagner	-13.6	-25.6	-47.2	21.3	9.8	33.9	29.0	75.2	476.0	163.5	6264	16	1183	June 8	Sep. 13	97
Whitecourt	-10.8	-22.3	-50.0	22.2	7.9	33.9	29.3	101.6	552.5	173.7	6151	15	1152	June 17	Aug. 25	69
Victoria	6.1	2.1	-15.6	19.7	11.1	35.6	110.7	13.4	647.2	32.0	2947	17	1937	Mar. 3	Dec. 5	276
Vancouver	5.1	0.6	-18.3	20.7	13.2	35.0	172.7	37.0	1257.7	54.7	3024	39	1964	Mar. 16	Nov. 16	244
Regina	-12.6	-23.2	-50.0	26.1	11.7	43.9	16.6	53.3	384.0	115.7	5877	137	1677	May 24	Sep. 11	109
Saskatoon	-14.1	-24.3	-47.8	25.4	11.5	40.0	17.8	54.2	348.8	113.1	6063	111	1620	May 21	Sep. 16	117
Winnipeg	-14.3	-24.2	-45.0	25.9	13.3	40.3	21.3	75.9	525.5	125.5	5923	178	1785	May 23	Sep. 22	121
Thunder Bay	-9.4	-21.3	-41.1	24.3	10.9	40.3	40.9	75.4	711.8	213.0	5768	66	1425	May 30	Sep. 12	104
Sudbury	-6.9	-17.8	-43.9	25.3	14.2	38.3	59.1	73.8	794.1	195.6	5043	143	1808	May 12	Oct. 8	142
Windsor	-1.2	-8.5	-26.1	27.7	16.7	38.3	55.0	83.4	848.8	117.4	3622	391	2533	Apr. 26	Oct. 21	177
London	-2.7	-10.5	-31.7	26.4	14.2	36.7	75.2	72.4	909.4	208.8	4133	237	2139	May 10	Oct. 5	147
Toronto	-1.3	-7.9	-32.8	26.7	17.2	40.6	60.9	74.0	800.5	139.2	3646	347	2440	Apr. 20	Oct. 29	191
Ottawa	-6.4	-15.2	-38.9	26.2	14.9	37.8	55.2	85.2	846.2	205.5	4634	235	2072	May 8	Sep. 30	144
Montréal	-5.2	-12.1	-37.8	26.1	17.4	37.8	74.0	95.3	1020.1	242.8	4198	327	2328	Apr. 19	Oct. 14	177
Québec	-7.5	-16.6	-36.1	24.9	13.2	35.6	89.8	116.6	1174.0	343.4	5165	123	1690	May 13	Sep. 28	137
Halifax	-0.1	-8.1	-26.1	21.7	13.0	34.4	143.3	97.1	1361.4	198.6	4186	56	1679	May 5	Oct. 23	170
St. John's	-0.5	-7.2	-25.6	20.2	10.7	31.5	155.8	75.3	1513.6	359.4	4824	29	1196	June 1	Oct. 11	131

APPENDIX 7
Mean percentage of time in January that the wind chill exceeds various limits at selected Canadian cities

Wind Chill	(Watts / m²)					
	1000	1200	1400	1600	1900	2300
EDMONTON						
Municipal Airport	94	78	62	45	13	*
Namao Airport	97	86	68	50	18	1
Calgary	89	72	52	35	10	*
Cold Lake	98	86	71	53	15	*
Coronation	97	86	68	51	17	1
Fort McMurray	98	89	73	57	15	*
Fort Vermilion	98	91	79	51	18	*
Grand Prairie	96	85	64	44	14	*
Lac La Biche	95	82	67	47	11	0
Lethbridge	92	72	48	28	4	0
Medicine Hat	94	75	53	32	5	*
Red Deer	96	84	63	44	16	1
Rocky Mountain House	85	65	47	29	4	0
Vermilion	96	84	69	52	17	*
Wagner	96	86	73	58	26	1
Whitecourt	93	78	60	41	11	*
Victoria	29	6	1	*	0	0
Vancouver	24	5	*	0	0	0
Regina	99	93	80	62	31	3
Saskatoon	99	93	78	63	29	3
Winnipeg	99	94	83	66	25	1
Thunder Bay	96	81	57	36	11	1
Sudbury	99	88	61	37	10	*
Windsor	81	51	23	8	*	0
London	88	59	28	11	1	0
Toronto	84	56	28	10	1	0
Ottawa	94	73	44	22	4	0
Montréal	92	70	40	18	3	*
Québec	95	76	49	26	6	*
Halifax	84	48	18	6	*	0
St. John's	91	62	23	6	*	0

* Less than 1

Class	Wind Chill watts/m²
Work and travel become uncomfortable unless properly clothed.	1000
Conditions no longer pleasant for outdoor activities on overcast days.	1200
Conditions no longer pleasant for outdoor activities on sunny days. Work and travel become more hazardous unless properly clothed. Heavy outer clothing necessary.	1400
Unprotected skin will freeze with direct exposure over prolonged period. Heavy outer clothing becomes mandatory.	1600
Unprotected skin can freeze in one minute with direct exposure. Multiple layers of clothing mandatory. Adequate face protection becomes important. Work and travel alone not advisable.	1900
Conditions for outdoor travel such as walking become dangerous. Exposed areas of the face freeze in less than 1 minute	2300

APPENDIX 8
Mean percentage of time that uncomfortably hot, humid conditions
exist at mid-summer at selected Canadian cities

Time	Percentage of days with one hour or more with humidex greater than:		Percentage of hours with humidex greater than:	
	30 (some people uncomfortable)	40 (everyone uncomfortable)	30 (some people uncomfortable)	40 (everyone uncomfortable)
EDMONTON				
Municipal	9	*	2	*
International Airport	10	0	2	0
Namao Airport	8	0	2	0
Calgary	3	0	1	0
Cold Lake	9	*	2	*
Coronation	11	*	2	*
Fort Chipewyan	6	*	1	*
Fort McMurray	10	*	2	*
High Level	6	*	1	*
Lethbridge	14	*	3	*
Medicine Hat	23	1	5	*
Peace River	6	0	1	0
Red Deer	9	0	2	0
Rocky Mountain House	7	*	1	*
Vermilion	11	*	2	*
Wagner	4	*	9	*
Whitecourt	7	0	1	0
Victoria	4	*	1	*
Vancouver	4	*	1	*
Regina	24	3	6	*
Saskatoon	18	2	4	*
Winnipeg	38	3	13	*
Thunder Bay	21	2	5	*
Sudbury	23	1	7	*
Windsor	63	7	28	1
London	50	3	19	*
Toronto	43	6	16	1
Ottawa	42	3	16	1
Montréal	43	5	16	1
Québec	28	3	9	1
Halifax	11	*	2	*
St. John's	8	0	1	0

* Less than 1

APPENDIX 9
Recreation season lengths for selected Canadian cities

	MEAN DATES					LENGTH (DAYS)				
	Winter Begins	Winter Ends	Spring shoulder begins	High summer begins	High summer ends	Winter	Spring shoulder	High summer	Autumn shoulder	Complete summer
EDMONTON										
Municipal Airport	Oct. 27	Apr. 18	May 3	May 22	Sep. 6	174	19	108	50	177
International Airport	Nov. 4	Apr. 23	May 8	May 22	Sep. 6	171	14	108	58	180
Namao Airport	Oct. 31	Apr. 21	May 6	May 24	Sep. 3	173	18	103	57	178
Calgary	Oct. 24	Apr. 28	May 13	May 31	Sep. 11	187	18	104	42	164
Cold Lake	Nov. 4	Apr. 17	May 2	May 24	Sep. 1	165	22	101	63	186
Coronation	Oct. 27	Apr. 19	May 4	May 25	Sep. 10	175	21	109	46	176
Fort Chipewyan	Oct. 24	Apr. 24	May 9	June 3	Aug. 26	183	25	85	58	168
Fort McMurray	Oct. 14	Apr. 21	May 6	May 23	Aug. 31	190	17	101	43	161
High Level	Oct. 29	Apr. 17	May 2	May 24	Aug. 30	171	22	99	59	180
Lethbridge	Oct. 27	Apr. 17	May 2	May 17	Sep. 24	173	15	131	32	178
Medicine Hat	Oct. 28	Apr. 13	Apr. 28	May 10	Sep. 27	168	12	141	30	183
Peace River	Oct. 13	Apr. 16	May 1	May 26	Aug. 31	186	25	98	42	165
Red Deer	Oct. 26	Apr. 22	May 7	May 27	Sep. 9	179	20	106	46	172
Rocky Mountain House	Oct. 13	Apr. 29	May 14	June 4	Sep. 5	199	21	94	37	152
Vermilion	Oct. 29	Apr. 14	Apr. 29	May 20	Sep. 6	168	21	110	52	183
Wagner	Oct. 23	Apr. 24	May 9	June 9	Aug. 28	184	31	81	55	167
Whitecourt	Oct. 12	Apr. 24	May 9	May 27	Sep. 2	195	18	99	39	156
Victoria	Dec. 25	Jan. 24	Feb. 8	June 19	Sep. 15	31	131	89	100	320
Vancouver	Dec. 22	Feb. 15	Mar. 2	June 14	Sep. 9	56	104	88	103	295
Regina	Nov. 1	Apr. 18	May 3	May 14	Sep. 18	169	11	128	43	182
Saskatoon	Oct. 27	Apr. 16	May 1	May 15	Sep. 13	172	14	122	43	179
Winnipeg	Nov. 7	Apr. 16	May 1	May 15	Sep. 17	161	14	126	50	190
Thunder Bay	Nov. 10	Apr. 15	Apr. 30	May 30	Sep. 9	157	30	103	61	194
Sudbury	Nov. 8	Apr. 20	May 5	May 18	Sep. 17	164	13	123	51	187
Windsor	Nov. 27	Mar. 26	Apr. 10	May 6	Oct. 6	120	26	154	51	231
London	Nov. 12	Apr. 1	Apr. 16	May 13	Sep. 28	141	27	139	44	210
Toronto	Nov. 26	Mar. 30	Apr. 14	May 14	Sep. 30	125	30	140	56	226
Ottawa	Nov. 13	Apr. 9	Apr. 24	May 13	Sep. 23	148	19	134	50	203
Montréal	Nov. 17	Apr. 8	Apr. 23	May 12	Sep. 24	143	19	136	53	208
Québec	Nov. 12	Apr. 21	May 6	May 22	Sep. 14	161	16	116	58	190
Halifax	Nov. 30	Apr. 4	Apr. 19	June 12	Sep. 20	126	54	101	70	225
St. John's	Nov. 18	May 5	May 20	June 30	Aug. 27	169	41	59	82	182

APPENDIX 10
Mean monthly total number of hours with bright sunshine for selected Canadian cities

	Jan.	Feb.	Mar.	Apr.	May	June	July	Aug.	Sep.	Oct.	Nov.	Dec.	Annual
EDMONTON													
Municipal Airport	90	116	168	228	278	272	306	277	182	162	107	80	2264
International Airport	98	119	172	233	284	287	313	284	183	163	103	78	2315
Beaverledge	74	109	160	209	271	277	301	263	169	141	89	62	2126
Calgary	102	128	162	205	254	268	322	282	195	176	124	98	2314
Edson	83	116	154	204	245	254	281	246	163	151	93	66	2056
Fort Vermilion	70	110	175	235	282	290	301	262	156	124	63	39	2107
Lethbridge	95	123	167	198	263	284	345	299	214	175	117	90	2378
Medicine Hat	93	122	162	201	271	279	348	298	199	173	112	87	2345
Victoria	68	96	151	202	277	275	342	288	206	145	83	59	2191
Vancouver	55	83	129	168	240	236	297	244	181	123	69	49	1872
Regina	100	121	156	209	278	283	342	295	191	168	104	84	2331
Saskatoon	103	135	190	232	292	294	341	293	191	175	109	86	2450
Winnipeg	121	144	176	220	266	276	316	283	185	152	91	93	2321
Thunder Bay	118	147	173	215	252	262	304	256	168	128	87	93	2203
Sudbury	101	132	152	207	247	246	288	251	151	122	78	85	2060
Harrow (Windsor)	83	104	123	169	237	253	278	254	188	162	89	72	2011
London	71	97	121	167	230	244	274	246	173	142	75	56	1894
Toronto	92	112	145	182	233	253	281	252	192	149	81	75	2045
Ottawa	99	120	148	177	239	247	274	243	168	136	80	79	2009
Montréal	96	114	157	179	234	246	270	236	177	137	78	80	2004
Québec	97	113	140	172	220	224	248	219	153	116	74	76	1852
Halifax	113	129	147	165	210	221	219	225	180	157	109	93	1969
St. John's	74	85	100	114	166	194	227	189	150	108	69	59	1532